T0093430

Applied Entomology:

Insect Ecology and Integrated Pest Management

Applied Entomology:

Insect Ecology and Integrated Pest Management

Lakshman Chandra Patel

Assistant Professor (Agricultural Entomology)
College of Agriculture
(Extended Campus of Bidhan Chandra Krishi Viswavidyalaya)
Burdwan - 713101, West Bengal

CRC Press
Taylor & Francis Group
Boca Raton London New York

CRC Press is an imprint of the
Taylor & Francis Group, an **informa** business

Elite Publishing House

First published 2024
by CRC Press
4 Park Square, Milton Park, Abingdon, Oxon, OX14 4RN

and by CRC Press
2385 NW Executive Center Drive, Suite 320, Boca Raton FL 33431

© 2024 Elite Publishing House

CRC Press is an imprint of Informa UK Limited

British Library Cataloguing-in-Publication Data
A catalogue record for this book is available from the British Library

Print edition not for sale in India.

ISBN: 9781032627786 (hbk)
ISBN: 9781032627793 (pbk)
ISBN: 9781032627809 (ebk)

DOI: 10.4324/9781032627809

Typeset in Adobe Caslon Pro
by Elite Publishing House, Delhi

–EPH–

Contents

About the Author

Dr. L.C. Patel is working as Assistant Professor (Entomology) in BCKV, Burdwan Sadar, West Bengal since 2014. Here, he also served as TIC for 6 months as additional responsibility. Prior to join BCKV, he served as Programme Coordinator in Khowai KVK, Tripura for 2 years 9 months and as SMS (Plant Protection) in S-24 Parganas KVK, W.B. for 4 years 4 months.

His academics is endorsed with graduation (Ag.) from UBKV, M. Sc. and Ph. D. in Entomology from BCKV & PGDAEM from MANAGE. He has expertise on IPM with special emphasis on biological control of insect pests.

He has various publications including 40 research papers, 3 books, 4 book chapters and 45 others. Dr. Patel has exposed with 26 scientific events and 20 professional trainings at regular interval to update his knowledge. As a resource person, he has successfully imparted 140 trainings (Practicing Farmers - 110, Rural Youth – 12 and Extension Functionaries – 18), 26 lectures , 28 TV programmes, 9 Radio programmes, 12 Farmers' Field Schools, 5 Farm schools along with organizing and active participating in number of different other extension activities like Field Day, Kisan Mela, Kisan Ghosthi, SHG & UG meet, Exhibition, Film Show, Method Demonstrations, Advisory Services, Participatory Rural Appraisal (PRA) programme etc.

He developed a Bio-Control and Plant Health Clinic Laboratory for Nimpith KVK during 2009-10 funded by NHM, GoI. He acted as West Bengal State Coordinator of IRM of Cotton Pests. He performed as Co-PI of NICRA project for West Tripura district and DBT project on Biotechnology Led Organic Farming in NEH region. He was engaged as nodal officer of IWMP batch IV – Khowai district, Tripura. He is now engaged in teaching. Some related national universities are utilizing his expertise as paper setter and external examiner. At present establishment, he acted as PI of some adhoc projects (14) funded by different sponsors. He is also involved as editorial and review member for some reputed Journals. Dr. Patel is associated as life member with 11 scientific societies.

He received 'Young Scientist Award' from SPPL, New Delhi (2015), CWSS, BCKV (2016) and AEDS, U.P. (2020). He also bagged 'Best Oral Presentation

Award' on Global Conference organized by AESSRA, New Delhi (2020); Fellow award (2020) from SBER, Tripura; Best Assistant Professor Award (2020) from Pearl Foundation, T.N., and Outstanding Achievement Award (2021) from STA, New Delhi. He had significant contribution to bag Zone II best KVK award - 2011 for Nimpith KVK.

Chapter - 1

Study of Distribution Pattern of Insects in Crop Ecosystem

Introduction

Knowledge of habitat-specific distribution patterns of a given insect population are of high ecological value. In agro-ecosystem, insects show various pattern of distribution on the basis of their taxonomic position, ecological role, ecological habitat etc. The dispersion of species is influenced by social instinct such as breeding, protection against natural enemies and heterogeneity of the environment. Individuals of a population arrange themselves in a manner that is specific to each population and these arrangements in space appear to be of considerable importance in the study of dynamics of ecosystem.

Objective

1. To validate methods for population estimation.

2. To understand the pest population.

3. To develop a sound sampling plan.

4. To gather information about behavior of the species.

Dispersion

The manner in which members of pest population are distributed in space is the dispersion or the distribution pattern of the species. Individuals in any population may be distributed according to three basic patterns.

1. Regular / Uniform distribution

2. Random/ Poison distribution

3. Clumped / Aggregated / Over-dispersed / Contagious distribution

To understand the distribution/dispersion of insect species one need to calculate,

A) Mean $(\bar{x}) = \sum fx/n$

B) Variance $(S^2) = [\sum fx^2 - (\sum fx)^2/n] / n\text{-}1$

Where, f= Frequency of number of plants/branches, x = Number of insects per plant/branch, n= Total number of plants

C) Variance Mean ratio (VMR) = $S^2/\bar{x}$

D) Index of David and Moore (IDM) =VMR - 1

E) Index of Lexis = $\sqrt{(S^2/\bar{x})}$

F) Charlier Coefficient = $\sqrt{(s^2 - \bar{x})} / x$

Distribution pattern

1. **Regular / Uniform distribution:** It may occur where competition between the insects is severe due to physical factors. Here,

   » The variance is less than Mean $(S^2 < \bar{x})$

   » Hence, the Variance Mean Ratio (VMR) is less than one,

   » IDM is less than zero,

   » Index of Lexis is more than one and

   » The Charlier Coefficient is less than zero.

2. **Random or Poison distribution:** It is relatively rare in nature and occurs where the environment is very uniform and there is no tendency to aggregate. Each insect has equal probability of occupying any point in space and the presence of one individual does not influence the distribution of another. Here,

» The Variance Mean Ratio (VMR) is always one ($S^2 = \bar{x}$)

» IDM is equal to zero,

» Index of Lexis is unity

» The Charlier Coefficient is zero.

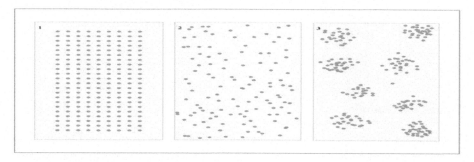

      Uniform   Random   Aggregated

3. **Aggregated/Clumped / Aggregate/Negative binomial distribution:** This is the most frequently observed pattern and individuals show varying degree of clumping together due to attraction or instinct as in case of some insects. Large scale clumping helps to evade possible danger of predation, climate or diseases. Bees are able to exist in cold climate by increasing the input of heat among them thus modifying the environments. Usually the environment decides the degree of aggregation of clumped patterns. Here,

» The Variance Mean Ratio (VMR) is more than one ($S2 > x$),

» IDM is more than zero,

» Index of Lexis is more than one

» The Charlier Coefficient is more than zero.

The aggregation may also be due to,

1. Response of microclimatic differences and daily and seasonal changes in weather.

2. As a result of reproduction or social attraction

3. Characteristics of the species *i.e.* degree of sociality

Aggregation always leads to intra-specific competition for food, space and reproduction etc. The degree of aggregation as well as the overall density, which results in optimum population growth and survival, varies with species and conditions.

Distribution Patterns	Uniform $(S^2 < \bar{x})$	Random $(S^2 = \bar{x})$	Aggregated $(S^2 > x)$
VMR	< 1	=1	>1
IDM	<0	=0	>0
IL	>1	=1	>1
CC	<0	=0	>0

Worked out example:

Workout the mean, variance and different indices to conclude the Distribution/Dispersion patterns of *Spodoptera litura* (Fab.) based on the following larval population data on cabbage.

Number of insects per plant (x)	Plant frequency (f)	fx	fx²
0	22	0	0
1	3	3	3
2	5	10	20
3	8	24	72
4	12	48	192
5	11	55	275
6	9	54	324
7	23	161	1127
8	7	56	448
9	6	54	486
10	4	40	400
Total	**n=110**	**$\sum$fx=505**	**$\sum$fx²=3347**

So, the mean, $(\bar{x}) = \sum fx /n = 505/110 = 4.6$

Variance $(S^2) = [\sum fx^2 - \{(\sum fx)^2 / n\}] / n-1 = [3347 - \{(505)^2 / 110\}] / 109 = 9.436$

Index of David and Moore (IDM) $= (S^2 / \bar{x}) - 1 = (9.436 / 4.6) - 1 = 1.05$

Index of Lexis $= \sqrt{(S^2/\bar{x})} = \sqrt{(9.436 / 4.6)} = 1.432$

Charlier Coefficient $= \sqrt{(s^2 - \bar{x})} / \bar{x} = \sqrt{(9.436 - 4.6)} / 4.6 = 0.48$

Conclusion:

Calculated IDM and Index of Lewis is both greater than 1 and Charlier Coefficient is greater than 0 but less than 1. Again, as variance > mean, it can be concluded that the distribution patterns of *Spodoptera litura* (Fab.) is Aggregated distribution.

Question-Answer

SAQs

1. What is the "Distribution pattern of insects"?

The manner in which members of pest population are distributed in space is the dispersion or the distribution pattern of the species. The internal distribution patterns are important which are related with some characteristics of a population

2. What are the objectives of to know the "Distribution pattern of insect in the crop ecosystem"?

i) Validate methods for population estimation

ii) Understand the pest population

3. What are the factors of distribution?

Breeding, protection against natural enemies and heterogeneity of the environment.

4. What are the main causes of distribution of insects?

There are several causes, but the main points are –

• Natural biogeography. • Crop Distribution. • Agricultural Practices (Monocultures, irrigation, fertilizers, pesticides) • Climate • Trade • Cultural Pattern etc.

5. How does the distribution of insects effect the crop ecosystem?

Insects are crucial components of crop ecosystem, where they perform many important functions. They aerate the soil, pollinate blossoms and control plant pests. Many insects, especially beetles are scavengers, feeding on dead animals and fallen trees, thereby recycling nutrients back into the soil. So, they play a major role in crop ecosystem. As decomposers, insects help to create top soil, the nutrient-rich layer of soil that helps plants to grow. Burrowing bugs, such as ants and beetles, dig tunnels that provide channels for water, benefiting plants. Bees, wasps, butterflies and ants pollinate flowering plants. The big-eyed bug and praying mantis control the size of certain insect populations, such as aphids and caterpillars, which feed on new plant growth. Finally, all insects fertilize the soil with the nutrients from their droppings. As, insects are one of the most required things in the crop ecosystem, so, their distribution on large scale is also important for the outcome.

6. How does the environment influence the insects to distribute in the environment?

Some insects tolerate cold shock but some fail to survive in cold. Insects show their wide adaptability to warm and hot climate, so dispersion of insects take place.

7. What are the types of insect distribution pattern?

1. Regular or uniform distribution pattern.

2. Random or poison type

3. Clumped or aggregated type of distribution pattern.

8. What is Regular distribution pattern?

When competition between the insects is severe due to physical factors, then it is called regular pattern of distribution.

9. What is random distribution pattern?

It is relatively rare in nature and occurs where the environment is very uniform and there is no tendency to aggregate. Each insect has equal probability of occupying any point in space and the presence of one individual does not influence the distribution of another.

10. What is aggregated distribution pattern?

This is the most frequently observed pattern and individuals show varying degree of clumping together due to attraction or instinct as in case of some insects.

11. What is the importance of VMR ratio in insect distribution pattern?

Variance/mean ratio (VMR) is used to characterize the distribution of events or objects in time or space.

12. What is the objective of studying distribution patterns of insects in crop ecosystems?

Study of distribution patterns of insects in a crop-ecosystem is most important to validate methods for population estimation and to understand the pest population

13. How the dispersion of species is influenced by social instinct?

The dispersion of species is influenced by social instinct such as breeding, protection against natural enemies and heterogeneity of the environment.

14. How the aggregated distribution pattern helps social insects?

Social insects like bees (*Apis spp.*) are able to exist in cold climate by increasing the input of heat among them thus modifying the environments

15. Mention one reason of aggregation distribution pattern?

As a result of reproduction or social attraction, insects go through aggregated distribution pattern.

16. The dispersion of a species is influenced by which factors?

Dispersion of species is influenced by social instinct such as breeding, protection against natural enemies and heterogeneity of the environment.

17. In which type of distribution/dispersion patterns individuals are evenly spaced?

In regular dispersion individuals are evenly spaced.

18. There are how many types of non-random distribution patterns?

Non-random distribution patterns are of two types: Regular and Clumped.

19. In which type of distribution patterns population members must be distributed independently of all other population members?

If dispersion is random, population members must be distributed independently of all other population members.

20. Which is the most common type of population dispersion?

Clumped distribution is the most common type of population dispersion.

21. How to calculate the variance of Insect species?

$$= [\Sigma fx^2 - (\Sigma fx)^2 /n]/n-1$$

22. How to calculate Charlier Coefficient?

$$= [\sqrt{S^2 - x}]/x$$

23: What are the characteristics of regular distribution?

The variance-mean ratio (VMR) is less than one, IDM is less than zero, Index of Lexis is more than one and the Charlier Coefficient is less than zero.

24: What are the characteristics of random distribution?

The variance-mean ratio (VMR) is always one ($S^2 = x$), IDM is equal to zero, Index of Lexis is unity and the Charlier Coefficient is zero

25: What are the characteristics of aggregated binomial distribution?

The variance-mean ratio (VMR) is more than one ($S^2 > x$), IDM is more than zero, Index of Lexis is more than one and the Charlier Coefficient is more than zero.

26: What is the cause of aggregation?

The aggregation may be due to

> 1. Response of microclimatic differences and daily and seasonal changes in weather.
>
> 2. As a result of reproduction or social attraction
>
> 3. Characteristics of the species i.e. degree of sociality

27: What is the degree of aggregation?

The degree of aggregation as well as the overall density, which results in optimum population growth and survival, varies with species and conditions.

28. Why random distribution is very rare in nature?

Random distribution occurs only where the environment is very uniform and there is no tendency to aggregate. So, it is relatively rare in nature.

29. What is VMR?

VMR is variance-mean ratio that is S^2/X

30. What is IDM?

IDM is Index of David and Moore, which is $S^2/X - 1$

31. In which distribution, variance is equal to mean?

In random/poison distribution, variance is equal to mean.

32. What is important in the study of dynamics of ecosystem?

Individuals of a population arrange themselves in a manner that is specific to each population and such arrangement in space appear to be of considerable importance in the study of dynamics of ecosystem.

33. What are the advantages and disadvantages of aggregated or clumped distribution?

Advantages - Large scale clumping helps to evade possible danger of -i) Predation. ii) Climate. E.g.-Bees are able to exist in cold climate by increasing the input of heat among them by extreme clumping. iii) Diseases.

Disadvantages - Aggregation always leads to i) Intraspecific competition for food ii) Space iii) Reproduction, mating etc. The degree of aggregation or the overall density varies with species and condition

34. For uniform distribution describe the VMR, IDM, IL, CI value.

VMR<1, IDM<0, IL>1, CI>0 .

35. For random distribution describe the VMR, IDM, IL, CI value.

VMR=1, IDM=0, IL=1, CI=0 .

36. For aggregated distribution describe the VMR, IDM, IL, CI value.

VMR>1, IDM>0, IL>1, CI>0 .

37. What are the relation between S^2 and X in uniform, random, and aggregated distribution?

In uniform distribution, $S^2 < X$

In random distribution, $S^2 = X$

In aggregated distribution, $S^2 > X$

38. How VMR and IDM values are calculated?

For the calculation of VMR and IDM value, we take the help of variance (S^2) and mean (X). VMR can be calculated by, VMR = (S^2 /X), IDM can be calculated by, IDM = (S^2 /X – 1).

39. Which distribution is known as least common type distribution in nature?

Random distribution is known as least common type.

40. Which distribution is as unpredictable distribution?

Random distribution.

41. Which distribution is affected by allelopathy?

Uniform distribution is affected by allelopathy.

42. What is the main difference between distribution and dispersion?

The difference between the dispersal and distribution is the population distribution refers to the area where an entire species occupies, while dispersion is only one subpopulation.

43. How does dispersal effect insect distribution?

Dispersal, or the movement and subsequent breeding of individuals from one area to another, strongly influences the population dynamics of a species. Local dispersal by insects may be effected by migratory behaviours, or by short host-seeking flights.

Fill in the Blanks

1. In random distribution pattern the VMR is

Ans = 1

2. In regular pattern the VMR is

Ans = <1

3. In aggregated pattern the VMR is

Ans= >1

4. Clumped distribution occurs where resources are-

Ans. Patchy

5. Individuals are arranged without any apparent pattern in

Ans. Random pattern.

6. The Index of Lexis in Rndom pattern is

Ans. 1

7. Aggregation always leads to_____ competition for food, space and reproduction etc.

Ans. Intraspecific

8. The variance-mean ratio (VMR) is always _____ in random distribution pattern.

Ans. one

9. _____ distribution pattern is quite rare in nature.

Ans. Random or poison

10. The Charlier Coefficient is _____ in random distribution pattern.

Ans. Zero

11. Large scale clumping helps to evade possible danger of _____

Ans. predation, climate or diseases

12. A truly random distribution may occur in more _____ communities.

Ans: mature.

13. Organisms can be distributed randomly or nonrandomly throughout their _____

Ans: habitat.

14. _____ suggests extensive interaction among individuals and/or other components of the environment

Ans: Clumping.

15. The dispersion of species is influenced by social instinct such as …………

Ans. breeding.

16. In regular pattern the IDM is ....

Ans: 0

17. The Index of Lexis in Random pattern is ....

Ans: 1

18. The Charlier coefficient in aggregated pattern is ....

Ans: > 0

19. .... distribution is characterized by the lack of any strong social interactions between species.

Ans. Random

20. ....is found in populations in which the distance between neighbouring individuals is maximized.

Ans. Uniform distribution

21. Highly territorial species exhibit.......... pattern, in which individuals are spaced at relatively equal distances from one another.

Ans. Uniform distribution.

22. Species that are highly tied to particular resources, such as food or shelter, tend to concentrate around those resources, and thus exhibit ....distribution.

Ans. Clumped distribution

23. Organisms unaffected by the placement of resources or other individuals exhibit ....distribution pattern.

Ans. Random

MCQs

1. Which type of dispersion is rare in nature?

A) Random   B) Clumped   C) Regular

Ans: A) Random.

2. What are the other names of regular distributions?

A) Uniform   B) Even   C) Over dispersion   D) all of the above

Ans: D) all of the above

3. What are the other names of aggregated type of distribution?

A) Clumped   B) Contagious C) Under dispersion   D) all of the above

Ans: D) all of the above

4. Which is most frequently observed pattern?

A) Random   B) Clumped   C) Regular

Ans: B) Clumped

5. In which pattern the interaction is weak?

A) Random   B) clumped   C) Regular

Ans: A) Random

6. If the Variance of the insect population in one field is more than the mean, the distribution pattern is – a) random / b) uniform / c) aggregated

Ans. c) aggregated

7. Index of David and Moore (IDM) is equal to zero in which type of distribution pattern – a) random / b) uniform / c) aggregated

Ans. b) random

8. Which is the most frequently observed pattern – a) random / b) uniform / c) aggregated

Ans. c) aggregated

9. Root over product of Variance-mean ratio gives – a) Index of Lewis / b) Index of David and / c) Charlier Coefficient

Ans. a) Index of Lewis

10. Variance - Mean ratio (VMR) is equal to 1 in which type of distribution pattern – a) Random / b) Uniform / c) Aggregated

Ans. a) Random

11. Study of distribution patterns of insects in a crop ecosystem is most important to (i)Validate methods for population estimation (ii) Understand the pest population (iii)both (iv) none of the above

Ans: (iii) both

12. Information on distribution or dispersion of pest species provides a valid base for (i)Developing a sound sampling plan (ii)Gives information about behavior of species (iii) both (iv) none of the above

Ans: (iii) both

13. There are how many types of distribution or dispersion patterns of insects? (i) 2 (ii) 3 (iii) 4 (iv)none of the above

Ans: (ii) 3

14. In regular distribution i) $S^2 < X$ iii) $S^2 > X$ iii) $S^2 = X$ iv) None of these

Ans- i) $S^2 < X$

15. In random or poison distribution Charlier coefficient is

i) < 0 ii) $= 0$ iii) > 0 iv) None

Ans- ii) $= 0$

True/False

1. Climate is one type of distribution factor.
Ans: True

2. In aggregated pattern insects are evenly distributed.
Ans: False

3. In random pattern insects have equal probability for occupying space.
Ans: True

4. Matting tends to distribute the insects in the environment.
Ans: True

5. The variance is less than Mean ($S^2 < \bar{x}$) in Random distribution pattern.

Ans. False

6. Index of David and Moore (IDM) = Variance-mean ratio (VMR) -1

Ans. True

7. In uniform distribution pattern each insect has equal probability of occupying any point in space.

Ans. False

8. Response of microclimatic differences and daily and seasonal changes in weather is one reason of aggregated distribution pattern.

Ans. True

9. Information on distribution or dispersion of pest species provides a valid base for gathering information about behaviour of the species.

Ans. True

10. Individuals exhibit no patterns of attraction or avoidance to any component of their environment.

Ans: True

11. Migration by insects is not an adaptation to a rapidly changing environment.

Ans: False

Chapter - 2

Sampling Techniques for the Estimation of Insect Population and Damage

Sampling

It is not possible or even desirable to count all insects in a habitat. Therefore, to estimate the population density of a pest or the damage caused by it to the crop, one has to resort to sampling. Randomization and choice of sample units are the fundamentals of sampling.

Objectives of sampling

> » To detect the presence or absence of pests
> » Quantify abundance of pests and their natural enemies
> » Follow the progress of an arthropod population through time by regular, periodic sampling.
> » The goal of monitoring is to reach a decision as to whether, or when, a pest population requires control action.

Sampling unit

It is the representation of a habitat from which insect's accounts are to be made. The sampling unit may be a plant or a branch of a plant, certain number of leaves or fruiting bodies, a clamp, a micro plot of 1 m² or number of plants per metre row length etc.

Sample size

The number of sample units (subsamples) per sample.

- Differs with nature of pest and crop
- Larger sample size gives accurate results

Components of an insect sampling programme

1. Knowledge of pest and natural enemies
    a. Identification   b. Life cycle and biology   c. Injury caused
2. Action / Economic thresholds
3. Sampling / Monitoring / Scouting plan or programme
4. Sampling equipment supplies

Sampling techniques

1. Absolute sampling - To count all the pests occurring in a plot.
2. Relative sampling - To measure pest in terms of some values which can be compared over time and space e.g. Light trap catch, Pheromone trap

Sampling/ Monitoring Tools

Clipboard –Keep all the scouting forms and field maps in one place.

Pencils- Carry a spare

Field maps –Jot notes, location pest problems and record observations

Scouting forms –Record sampling monitoring data, field history.

Hand lens –See and correctly identify pests, 10-20x

Pocket knife –Cutting shoots, scraping at trunks skinning berries.

Shovel /sturdy trowel – Digging soil

Collection bags and vials –Send pest /damage samples to others

Traps/ trap parts (lures) – There's always a well functioned trap

Camera –Send pest /damage photo to others for ID

References – Field guides, fact sheets, pictures of pests/damage

GPS unit – Relocate sample sites accurately

Sampling techniques

1. **Active collecting** – It includes pooter, portable suction device, sweep net, visual observation etc.

2. **Passive collecting** – It includes coloured pan traps, emergence traps, sticky traps, suction traps, light traps, pit fall traps etc.

Methods of sampling

a. In situ counts –Direct visual counting or observation of number of insects on plant canopy (either entire plot or randomly selected plot). For example, counting of hoppers or eggs or larval population on leaves or plant surface.

b. Knock down - Collecting insects from an area by removing from crop and counting through following ways

Jarring: Collection of beetles or *Helicoverpa* larva on below placed cloth after jerking or beating on plant surface.

Chemicals: Plant is enclosed in a polythene envelope and treated with quick knockdown insecticide such as pyrethrum or dichlorvos. After sometime, the plants may be shaken to remove dead or immobilized insects on a cloth or other container for counting.

Heating: Heating forces the insects to come out of their dwellings by increasing the temperature of their habitat.

c. Netting - Use of sweep net for hoppers, dragonfly, grasshopper etc., vacuum net for leaf hoppers and aerial net for sampling air borne insects.

d. Trapping – Light trap - Phototropic insects, Bait trap –fruit fly, Suction trap – winged aphids and leaf hoppers, Water trap – brown plant hopper, Pitfall trap – ground moving insects, Pheromone trap - Species specific, Sticky trap - Sucking insects

Direct counting	Jarring	Netting
Light trap	Suction trap	Water trap
Pitfall trap	Pheromone trap	Sticky trap

e. Crop samples - Plant parts removed and pest counted

f. Fecal pellet collections - Lepidopterous larvae produce relatively large, solid, dark fecal pellets, most of which fall to the ground or lower leaves of the plant. The size, colour and freshness of the pellets indicate whether the caterpillars are young or mature and live or absent.

g. Mark, release and recapture method – Insects are captured from field, marked with suitable paint or dye, released in the field and recaptured along with unmarked insects. The population can be estimated by Lincoln index: $P = an/r$, where P = Population size; a = total of marked and released individuals; n = total number of individuals recaptured and r = the number of marked individuals recaptured

h. Indirect technique – Population is estimated by measuring the effect of insects on crop plants or some indirect sampling methods or population indices. For example, number of leaves mined by leaf miners, percent defoliation by hairy caterpillar, percent plant attack etc. Sometimes population is assessed with the help of insect products like larval or pupal skin, frass or honey dew etc.

i. Remote sensing – Radar can automatically provide information of aerial migration of pests and natural enemies. It can also detect large area damage on crop due to pest infestation.

Stage of Sampling

» Usually most injurious stage counted

» Sometimes egg masses counted - Practical considerations

» Hoppers - Nymphs and adult counted

Sampling and Monitoring of Natural Enemies

» Predators, such as ladybird beetle adult and larvae, syrphid fly larvae, lacewing larvae, and spiders.

» Evidence of parasitism, such as aphid "mummies," darkened greenhouse whitefly pupae, and scale insects with exit holes of the parasites.

» Signs of insect diseases, such as blackened, dead caterpillars and dead, discolored aphids infected with fungi.

Estimation/Assessment of insect population and damage

Need

a. To know the extent of pest load and their damage.

b. To workout economic injury level (EIL) and economic threshold level (ETL).

c. To estimate yield loss

d. To decide the timing of control measures in order to avoid indiscriminate use of insecticide.

Decision making

» Population or damage assessed from the crop

» Compared with ETL (Economic Threshold Level) and EIL (Economic Injury Level)

» When pest level crosses ETL, control measure has to be taken to prevent pest from reaching EIL.

Economic Injury Level (EIL): Cost of control measures = Loss by insect

$$\text{EIL} = \frac{C}{V \times I \times D \times K}$$

Where,

EIL = Economic injury level in insects/production (or) insects/ha

C = Cost of management activity per unit of production (Rs. /ha)

V = Market value per unit of yield or product (Rs. /tonne)

I = Crop injury per insect (Per cent defoliation/insect)

D = Damage or yield loss per unit of injury (Tonne loss/% defoliation)

K = Proportionate reduction in injury from pesticide use

Worked examples of EIL

Calculate EIL in terms of pest population/ha with following figures

C = Management cost per unit area = Rs.3, 000/- per ha

V = Market value in Rs. /unit product = Rs.1, 000/tonne

I = Crop injury/pest density = 1% defoliation/100 insects

D = Loss caused by unit injury = 0.05 tonne loss/1% defoliation

K = Proportionate reduction in injury from pesticide use= 0.8 (80% control)

$$EIL = \frac{C}{VIDK} = \frac{3000}{1000 \times 0.01 \times 0.05 \times 0.8}$$

EIL = 7500 insects/ha

Economic Threshold Level (ETL): Level at which, control measures to be taken to avoid the insect population / damage reaching EIL.ETL represents pest density lower than EIL to allow time for initiation of control measure

Factors Influencing ETL and EIL

Primary factors

a. Market value of crop - When crop value increases, EIL decreases and vice-versa

b. Management of injury per insect - When management costs increase, EIL also increases

Secondary factors

a) Degree of injury per insect

  » Insects damaging leaves or reproductive parts have different EIL (Lower EIL for reproductive part damages)

  » If insects are vectors of disease, EIL is very low even 1 or 2 insects if found - management to be taken

  » If insects found on fruits - Marketability reduced - EIL very low

b) Crop susceptibility to injury

  » If crop can tolerate the injury and give good yield. EIL can be fixed at a higher value

  » When crop is older, it can withstand high pest population - EIL can be high

Tertiary factors

Weather, soil factors, biotic factors and human social environment. These tertiary factors cause change in secondary factors thereby affect the ETL and EIL.

Pest Injury vs Damage

Injury – The effect that the pest has on the crop or commodity.

Damage – The effect that injury has on man's assessment of the crop's economic value.

For crops, "Injury" is biological and "Damage" is economic. For non-crops, "Injury" = "Damage".

Estimation of insect damages – It is useful

a. **Mechanical protection** – The crop yield from mechanically protected condition (Netting or other barriers to keep away the crop free from insect infestation) is compared with that obtained from infested crop

b. **Chemical protection** – The yield of treated plot is compared with that of untreated plot

c. **Comparing in different fields** – Yield is compared in different fields having different level of infestation. Correlation between crop yield and level of infestation is worked out to estimate the loss in yield

d. **Comparing yield of individual plant** – Yield of individual plant is measured in the same field and the average yield of healthy plant is compared with other plants having different degree of infestation and the loss is estimated.

e. **Damage caused by individual insect** – The details regarding the amount of damage caused by different stages of pest is worked out and loss is calculated

f. **Manipulation of natural enemies** – Pest is controlled by introduction of natural enemies and the yield is compared with the field where the natural enemies are not used.

g. **Simulation of damage** – Pest injury is simulated by removing or injuring the plant parts. The simulated damage may, however, not always be equivalent to the damage caused by an insect.

Assessment of damage

a. Rice

Sucking insects

1. Green Leaf hopper (GLH): Feeding on leaves results in yellowing. Vector for rice tungro virus disease (RTV). Count the number of insects per seedling in the nursery (ETL: 50/100 seedling) (or) number per hill in the field (ETL: 5/hill at vegetative stage, 10/hill at reproductive stage (or) 2/hill in RTV endemic area). Sweep net can also be used for sampling. (ETL: 60/25 sweeping).

2. Brown planthopper (BPH): Feeding on stem just above water level results in hopper burn. Count the total number of insects in 10 hills selected at random in one square meter area (ETL:5-10/tiller).

3. **Rice ghandhi bug:** Black spot at feeding point on the grain and individual chaffy grains. Insects emit stinky odour. Count the number of bugs in 100 ear heads selected at random (ETL: 5 (flowering stage) or 16 bugs (milky stage) / 100 panicles).

Chewing Insects

1. Rice stem borer: Based on eggs and larval damage: Presence of yellowish brown egg mass near the leaftip/presence of dead heart (vegetative stage) or white ear (reproductive stage).

a. Eggs in the nursery: Number of egg masses/m^2 (ETL: 2)

b. Larval damage: Count the total tillers and affected tillers in a unit area and arrive at a percentage.

$$\% \text{ dead heart} = \frac{\text{Number of dead hearts}}{\text{Total number of tillers}} \times 100 \text{ (ETL: 10\%)}$$

$$\% \text{ white ears} = \frac{\text{Number of white ears}}{\text{Total productive tillers}} \times 100 \text{ (ETL: 2\%)}$$

2. Leaf folders: Based on damage - folded and scrapped leaves

$$\% \text{ leaf damage} \frac{\text{Number of damaged leaves}}{\text{Total number of leaves (in 10 randomly selected)}} \times 100$$

(ETL: 10% at vegetative stage or 5% at flowering stage)

B. Vegetables

a) Brinjal

1. Shoot and fruit borer: At vegetative stage, attack the shoot resulting in drying and dropping of shoots. Count the total number to number of shoots damaged and arrives at a percentage.

Fruit: Number basis: Count the infested as well as healthy fruits and arrive at a percentage. **Weight basis:** Weigh the damaged and healthy fruits and workout the percentage at harvest.

b) Bhendi

1. Fruit borer: Count the total number and number of fruits damaged in a plot and arrive at a percentage.

2. Leafhopper:

a) Based on number of insects in three fully opened leaves, express as number/leaf.

b) Grading the damage visually by observing the leaves in sample plants.

Grade I: Free from hopper burn

Grade II: Crinkling and curling of a few leaves mostly in upper portion of the plant and yellowing.

Grade III: Crinkling and curling of leaves all over the plant and stunted growth.

Grade IV: Extreme crinkling and curling hopper burn, leaf shedding and stunted growth.

Question-Aanswer

MCQs

1. Which type of trap is used for aphids, flies?

a) Sticky trap   b) Light trap   c) Pitfall trap   d) nets

Ans: a) Sticky trap

2. By which method, moths and beetles are collected?

a) Suction trap   b) Light trap   c) Pitfall trap   d) Quadrat method

Ans: b) Light trap

3. Sampling method used to collect leafhopper is?

a) Suction trap b) Windowpane method   c) Vaccum netting   d) Sticky trap

Ans: c) Vaccum netting

4. In case of Pitfall trap, which type of insect is collected?

a) Ground-dwelling insects b) Flying insects c) Arboreal insects d) None of the above

Ans: a) Ground-dwelling insects

5. Which method is suitable for catching insects associated with lower vegetation like smaller trees and shrubs?

a) Malaise trap b) Shaking & Beating method   c) Emergence trap   d) Sticky trap

Ans: b) Shaking and Beating method

6. A Sample is a portion of the _____ population that is considered for study and analysis:

a) Selected   b) Total   c) Fixed   d) Random

Ans: b) Total

7. For a sample to be truly representative of the population, it must truly:

a) Fixed   b) Random   c) Specific   d) Casual

Ans: b) Random

8. Which of the following statements are true in regard of Sampling:

a) Smaller Sample more accuracy b) Large Sample more accuracy
c) Larger Sample less accuracy

Ans: b) Large sample more accuracy

9. Which of the following is not a mechanical method:

a) Hand picking   b) Shaking   c) Wrapping   d) Use of light traps

Ans: d) Use of light traps

10. Which one of these is not a sampling technique?

a) Suction   b) Netting   c) Trapping   d) Planting

Ans: (d) Planting

11. Counting insects on plant parts is which type of sampling method?

a) Chemical   b) Physical   c) Bio-chemical   d) Pheromonal

Ans: (b) Physical

12. Yellow Sticky Traps are mainly used for which insect?

a) Aphids   b) Whiteflies   c) Leaf miners   d) All of the above

Ans: (d) All of the above

13. Which component is not including in insect sampling?

a) Identification b) Sampling c) Monitoring d)Manipulation of natural enemies

Ans. d

14. Which method is widely used for the estimation of insect pest?

a. Light trap   b. Pheromone trap   c. Netting   d. All of the above

Ans: d

15. The common attractant used for attracting sorghum shoot fly

a) Mixture of yeast   b. Mixture of molasses   c. *Atherigona soccata*   d. None

Ans: d

16. Which is the sampling technique for estimation of crop yield damage?

a) Pheromone trap   b) Light trap   c) Chemical protection   d) Netting

Ans: c

17. Types of sampling programme are –

a) Two   b) Three   c) Four   d) Five

Ans. a) Two

18. The fundamentals of sampling are –

a) Choice of sample unit   b) Randomization
c) Randomization and choice of sample unit   d) Monitoring

Ans. c) Randomization and choice of sample unit

19. For the recording of egg or larval population of *Helicoverpa armigera*, the technique is used –

a) Knockdown   b) In situ count   c) Netting   d) Sieving

Ans. b) In situ count

20. The efficiency of flotation could be increased by adding some salt such as –

a) Magnesium sulphate   b) Calcium sulphate

c) Magnesium phosphate   d) Calcium phosphate

Ans. a) Magnesium sulphate

21. For sampling winged aphids and leafhoppers, more useful technique is–

a) Window trap   b) Malaise trap   c) Vacuum netting   d) Suction trap

Ans. d) Suction trap

SAQs

1. What are the tools of attractant traps?

Yellow sticky trap, pheromone trap etc.

2. Give two examples of relative method for catching and trapping insects.

Sweep-net catch, sticky traps.

3. Write two purposes for the estimation of insect population?

To determine local or newly introduced population and to monitor pest level to control the pest-related problems.

4. What is Quadrat method?

Quadrate method is an ecological sampling method consisting of a small square area of the ground within which all species of interest (especially insect pests) are noted or measurements taken.

5. What is absolute sampling?

It is a sampling method that estimates in density per init such as locust per land area, pupae   per tree.

6. What is the order of insects which are collected by malaise trap?

Hymenoptera and Diptera

7. Which trap is used to capture of the nocturnal insects?

Light traps

8. In which sampling technique pesticide is used?

Knockdown sampling.

9. What is pheromone traps?

Pheromone trap is one of the sampling methods that is used for the relative assessment of insect population. This trap usually attracts only one particular sex.

10. In which sampling method, fish can be used?

Bait trap.

11. What is Insect Sampling?

It is referred to as scouting or monitoring. Thus the main objectives of insect sampling are to detect what species are present. Determine their population density and how they are distributed in the field.

12. What are the importances of Insect Sampling?

Assessment of pest density usually requires obtaining actual counts of the pest, and therefore sampling is important. Because sampling is time consuming and expensive, one must know how to gather information about Pest abundance to be able to make correct decisions without incurring excessive costs.

13. Which types of shapes are commonly used for sampling?

"W" or "U" shaped sampling patterns are more commonly used in a square shaped field. In a long narrow field, a "zig-zag" or "Z" sampling pattern is usually more efficient.

14. What is the time of insect Sampling?

In general, it is a good idea to begin sampling as soon as the crop is transplanted into the ground or when plants emerge from the soil if direct seeded.

15. What is damage assessment?

The damage to a crop is another indicator of the number of pest insects present on the particular crop. Pest Management strategies are often based on the results of damage assessments; the application of an insecticide would be justified.

16. What is direct assessment?

Direct assessment aims at the assessment of a pest population causing particular damage. This is however in many cases not possible because the insects might be hidden in the plant, like termites.

17. What is indirect assessment?

It does not count the number of insect, but their representatives such as frass, the number of cocoons, exuviae, egg shells etc.

18. What is light trap?

Light traps, with or without ultraviolet Light, attract certain insects. Light traps are widely used to survey nocturnal moths.

19. What is 'injury' in IPM?

The effect that the pest has on the crop or commodity is known as injury.

20. What is 'damage' in IPM?

The effect that injury has on man's assessment of the crop's economic value is known as damage.

21. State a difference between injury and damage for crops.

For crops, "Injury" is biological and "Damage" is economic.

22. Name one primary factor influencing ETL and EIL.

'Market value of crop' is a primary factor influencing ETL and EIL.

23. What is Relative Sampling Technique?

Measuring pest in terms of some values which can be compared over time and space (e.g. Light trap catch, Pheromone trap is known as Relative Sampling Technique.

24. Name different sampling techniques.

The different sampling techniques are: Visual inspection, Knockdown, Suction, Netting, Trapping.

25. State one purpose of insect population estimation.

One main purpose of insect population estimation is : to determine a pest species, its population distribution, change in population in space & time.

26. What is the other name of Capture-recapture Method?

Capture-Recapture Method is also known as Peterson Lincoln index or proportionality method

27. What are used as sampling units ?

Traps, plant, leaf, berry are used as sampling units .

28. State some reason for sampling?

For making cost effective and environmentally sound insect management decisions & to avoid unnecessary treatments.

29. For which type of insects trapping methods are used for sampling?

Psyllids, certain aphids, plant bugs and spider mites

30. Which type of insects produce relatively large, solid, dark, fecal pellets used for sampling?

Lepidopterous larvae like catalpa caterpillars, oakworms, and datanas

31. Sticky band can be used for which insects?

Mango mealy bug, Walnut datana and spring canker worm adults.

32. State some Population Estimation Methods of Insects

Capture-Recapture Method, Relative Methods, Using a leaf Position

33. How beating samples are prepared?

A sampling tray is held horizontally just beneath plant foliage, and the foliage above is struck sharply a standard number of times (2 to 5) with a short stick or the other hand. Arthropods falling to the tray are immediately counted and then shaken off. This process is repeated several times around the periphery of the plant.

34. What are the relative methods for population estimation?

Visual searches, fixed time collection, sweep-net catch, shaking and beating, vacuum traps, malaise traps, window pane traps, sticky traps, pitfall traps and traps using attractants like pheromones.

35. What is sampling?

Sampling is a method to estimate the population density of a pest or the damage caused by it to the crop.

36. What are the components of insect sampling programme?

Knowledge of pest and beneficial insects, Action/ economic thresholds of pests, Identification, Monitoring plan or program, Life cycle and biology, Injury caused

37. Name some methods of sampling used for estimation of insect population.

Light Trap, Bait Trap, Pheromone Trap, Netting etc.

38. What are the target pests for pheromone traps?

Fruit flies (*Dacus spp.*), spotted bollworms (*Earias spp.*), cotton bollworm (*Helicoverpa armigera*), codling moth (*Cydia pomonella*), pink bollworm (*Pectinophora gossypiella*) and leaf eating caterpillar (*Spodoptera litura*).

39. What is the difference between malaise trap and suction trap?

The malaise trap is basically a tent made of cotton or nylon mesh with one side open that intercepts flying insect In Suction trap consists of a wire-gauze funnel leading to a collecting jar and a motor-driven fan is situated below the funnel to create the suction.

40. Write down different techniques adopted for the estimation of crop losses.

Chemical protection, Mechanical protection, Comparison of yield in different fields

41. What is sampling unit?

It is the proportion of habitat from which insects accounts are to be made. The sampling unit may be a plant or a branch of plant. Example: A certain number of leaves or fruiting bodies per hill or a clump or a micro plot of 1 square meter or number of plants per 1 meter row of length etc.

42. Name the different types of trapping system.

There are 9 different types of trapping systems. These are light trap, bait trap, pheromone trap, sticky trap, pit fall trap, malaise trap, suction trap, window trap and water trap.

43. Which is the most common method for collecting the white grub beetles?

Jarring

44. What is Lincoln index?

$$P = an/r$$

Where, P = population size, a = total number of marked and released individuals, n = total number of individuals recaptured and r = the number of marked individuals recaptured. This technique is useful for estimating population of mobile insects.

45. What is sample?

The individual units (subsample) from which insects are counted. The counting from one or more inspections at a scouting stop. All of the sample units (subsamples) collected to estimate the population density of pests or beneficial insects or mites in a field or portion of field.

46. Give advantages and disadvantages of pheromone trap.

Advantages: Pheromone traps are specifically attractive to the target species; there is no sorting or identification problem. Further, no power is required as in the case of light traps. So, these can be installed in any field.

Disadvantage: The major disadvantage with such traps is that almost all traps detect only the adult males.

47. What is remote sensing?

Remote sensing techniques used in entomology include photography and videography from aircraft and from the ground; satellite-borne photography, multispectral scanning, and thermal imaging; ground-based and airborne radar; and acoustic sounding and low-light optical methods. This could be particularly useful for pests which produce visible symptoms of crop damage over large area e.g., hopper burn symptoms in paddy.

48. What do you mean by sampling programme?

Sampling Programme is the procedure which employs the sampling technique to obtain a sample and make a density estimate. A sampling programme would envisage insect stage to be sampled, number of sampling units required, spatial pattern to obtain sampling units and timing of samples. It is of two types-Extensive and Intensive.

49. How many types of netting are used to trap the insects?

There are 3 types of netting which are usually used to trap the insects. These are – a) Sweep netting, b) Vacuum netting, c) Aerial netting.

50. What are the population estimation methods of insects?

1) Total counts of insect populations are in most cases labour intensive & time consuming

2) Parameter-estimating sampling or census or total counts gives an accurate estimate of a population

3) Decision-making sampling or monitoring or sample count allows precise estimates.

51. What is knockdown?

Knockdown is similar to in situ counting, except that the insects are dislodged from the plants causing them to fall into either a tray, a funnel or on a sheet or a piece of cloth and are subsequently counted. The method of dislodgement may be jarring, chemicals or heating.

52. What are the fundamental components of an IPM program?

Sampling and Monitoring (scouting) are the fundamental components of an IPM program.

Fill in the Blanks

1. Sampling unit is proportion of the _____ from which insect counts are to be made.

Ans: habitat

2. Estimation procedures provide the most information about the_____ being sampled.

Ans: population

3. The concept of Pest management was proposed in _____:

Ans: 1961

4. Plant protection in India and most of the developing countries is mainly based on the use of _____:

Ans: pesticidal chemicals

5. The full form of EIL is ………………………………..

Ans: Economic Injury Level.

6. The full form of ETL is ………………………………..

Ans: Economic Threshold Level.

7. One tertiary factor influencing EIL and ETL is …………………………….

Ans: Weather.

8. Sex pheromone traps mostly attract the .... of the species

Ans: males

9. Damage Assessment also helps in assessment of _____

Ans: yield loss

10. _____ is mainly used to attract adult insects into attractant traps.

Ans: Sex pheromones

11. _____ is a method of sampling used for estimation of insect population.

Ans: Light trap

12. _____ is a technique adopted for the estimation of crop losses by insect.

Ans : Comparison of yield in different fields

13. Remote sensing can be particularly useful for monitoring ....

Ans: Locust swarms.

14. The use of satellite images and .... allows an effective damage assessment of large-scale plantations.

Ans: aerial photos

15. Berlese funnels are more frequently used to sample small .... in the soil.

Ans: arthropods

16. .... traps are more useful in dry areas since in other areas rain may flood them and render them non-functional.

Ans: Pitfall

17. For monitoring sorghum shoot fly (*Atherigona soccata*) a .... trap is quite effective.

Ans: fish-meal

True or False

1. The damage caused by wood boring beetles can be assessed by counting the number of boreholes per area e.g 10cm × 10cm

Ans: True

2. Use of sterile male techniques of pest is a biological method

Ans: False

3. A sampling unit is proportion of the habitat from which insect counts are to be made.

Ans: True

4. Environmental awareness is proposed in 1970's

Ans: False

5. Sampling should be done at least once per week.

Ans: True

6. 'Crop susceptibility to injury' is a primary factor influencing ETL & EIL.

Ans: False

7. Larger sample size gives accurate results.

Ans: True

8. When sampling for pests, the person sampling should not look for predators.

Ans: False

9. GPS in sampling is used to relocate sample sites accurately.

Ans: True

10. The main reason for Sampling is to make cost effective and environmentally sound insect management decision.

Ans: True

11. The degree of defoliation of a host tree is a useful measure of the damage and is commonly indicated insect population.

Ans: True

12. Mark, release and recapture technique is useful for estimating populations of mobile insects.

Ans: True

13. Malaise trap is useful for monitoring ground dwelling predators.

Ans: False

14. Pheromone traps have been widely used in detecting or sampling fruit flies (*Dacus spp.*), spotted bollworms etc.

Ans: True

15. Extensive programmes are conducted as part of research in population ecology and dynamics.

Ans: False

Chapter - 3

Habit, Habitat, Distribution, Sampling and Identification of Mite Pests

Introduction

Order: Acarina, Subclass – Acari, Class: Arachnida, Phylum: Arthropoda

The mites that feed on plant are called phytophagous mites and belong to the families Tetranychidae (spider mites), Tenuipalpidae (false spider mites), Tarsonemidae (tarsonemids), Eriophyidae (blister or gall mites) and Eupodidae (eupodids). Of these spider mites are the most important and prevalent. Mites normally feed on the undersurface of the leaves but the symptoms are more easily seen on the upper surface. Tetranychids produce blotching on the leaf-surface, tarsonemids and eriophyids produce distortion, puckering or stunting of leaves and other parts of the plant. Some species of eriophyids produce distinct galls or blisters.

Family: Tetranychidae

1. *Tetranychus cinnabarinus, Tetranychus neocaledonicus, Tetranychus ludeni*:

These species have a world-wide distribution. Its infestation recorded on cotton, castor, pulses, groundnut, dhaincha, brinjal, cotton and bhendi are the worst sufferers.

Symptom of damage: Under surface of the leaves get covered with strands of webbing which affect photosynthesis and so the yield. Chlorotic spots coalesce into pale or silvery patches. Eventually the leaves dry up and fall off. Growth, flowering and fruit setting in the plants are greatly affected. Both nymphs and adults cause the damage.

Oligonychus indicus:

It is a serious pest of sorghum, maize, sugarcane and some cereals.

Symptom of damage: White or red patches on the lower surface of leaves of sorghum and sugarcane. Both nymphs and adults cause the damage.

Oligonychus oryzae:

It infests rice. White spots on lower surface of leaves which coalesce leading to development of white or silvery patches. It is severe during summer.

Family: Tarsonemidae

Polyphagotarsonemus latus **(yellow mite, broad mite, chilli muranai mite):**

These are pests of chilli, cowpea, greengram, horsegram, sesamum, lablab, jute and cotton.

Symptom of damage: Sudden downward curling and crinkling of leaves followed by development of blister patches. Severe stunting of growth and death of plants.

Family: Tenuipalpidae (False spider mites)

Brevipalpus lewisi

It is a non insect pest of citrus

Brevipalpus californicus

It causes serious injury to a wide variety of ornamental and agricultural crops.

Family Eriophyidae

Aceria cajani:

It is a vector of sterility mosaic of pigeonpea.

Aceria sorghii:

Leaf crinkling, general chlorosis and choking of terminal leaves. Host plant is sorghum.

Mites as vectors:

Some of the eriophid mites act as vectors of some important viral diseases Eg: *Aceria cajani* transmits redgram sterility mosaic disease. *Aceria tulipae* transmits wheat streak mosaic disease.

Mites as parasites: Ecto Parasite on honey bee: *Tropilaelaps clareae,* Endo Parasite: *Acarapis woodi* (Tarsonemidae,Tracheal mite on honey bee), *Locustacarus buchneri* (Poapolidae) on bumble bees

Predatoy mites: *Phytoseilus persimilis, Amblyseius fallacies*

Stored grain mite: *Acarus siro*

Distribution: Mites are large family with worldwide distribution. Most of the phytophagous mites and predatory plant mites occur in the field throughout the year in tropical climate except, of course, in rainy seasons when the population declines considerably due to washing away of the leaf population and also during the severe winter months when the egg laying ceases due to dropping down of temperature below the development threshold. Temperature, humidity and light are the important factors influencing the dynamism of mites.

Identification of mites: Specialized training is required for a novice to identify mites to species using taxonomic keys. Identification of mites, even to family level, requires learning to slide-mount mite specimens in the correct orientation so relevant morphological characters are visible. The tarsi, mouthparts, dorsal and ventral plates, setal patterns, and other setal structures should be observed carefully. Preservation of the mite specimen is done in ethyl alcohol (70-80%). A slide-mounting medium, such as Hoyer's mounting medium (25 g distilled water; 15 g gum arabic; 100 g of chloral hydrate, and 10 g of glycerol) is needed for mounting the specimen. Often, both adult males and females are required to key mites to the species level. Phase-contrast microscope or other high-quality compound microscope is used to see the structures on both the dorsum and venter of the mite.

Sampling and identification of mite pests are done to know if the mites are harming or disease causing to the plant population or not. Periodic collection, sampling and identification are done in an area to check the mite population density, introduction of new species so that pest management measures can be taken before spreading of any disease.

Damage by Spider mites on different crops

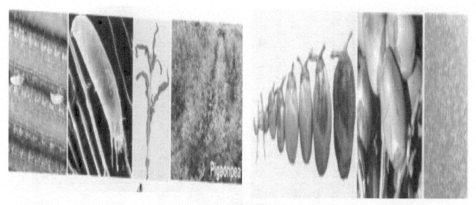

Damage by Eriophyid mite

Damage by Tenuipalpid mite

Damage by Broad mite

Identifying characteristics of mites:

Teranychidae	Tarsonemidae	Tenuipalpidae	Eriophidae
1. Body colour is red, green, yellow, brown etc. 2. Body is 0.2-0.8 mm long 3. Body is flat, oval. 4. Body is not divided into divisions 5. Not segmented. 6. Body of male tapers posteriorly 7. Chelicerae are fused to form a stylopore and the movable segment of chelicerae forms a flagellate stylet. 8. There is no mitotic division in larval stage. 9. Most of the species are having narrow host range 10. Palpal thumb claws are present.	1. Same as Tetranychidae, but without thumb claws. 2. Three types of setae namely hysterosomal, dorsocentral and mediolateral are present. 3. The true tarsal claw is hooked or pad like and with tenent hairs.	1. Body is elliptical (ovoid) 2. Body measures 0.1-0.3 mm long 3. Body is divided in to three parts Capitulum, Propodosoma and Prohysterosoma (the latter two parts together known as Idiosoma) 4. Mouth parts are contained in a distinct capsular head known as Capitulum 5. Females are bigger than males. 6. Body colour is opaque white, light green, pinkish 7. Adult integument is hard and shiny. 8. Few hairs, spines are present on body. 9. Chelicerae are needle like	1. Body is minute measuring 0.08 – 0.2 mm long 2. Body is 2 types: a) Elongate (vermiform), worm like, soft body b) Wedge shaped, hard body 3. Body is segmented 4. Body is divided into cephalothorax and tapering abdomen 5. Abdomen is finely striated with long setae 6. Two pairs of legs on anterior end of body (in all the life stages) 7. Pedipalpi or chelicerae are capable of making some independent movements and form a telescope or fold base. No thrusting stylopore.

Egg--Larva (3 pairs of legs) -- Protonymph (4 pairs of legs) – Deutonymph (4 pairs of legs) – Tritonymph (4 pairs of legs) -- Adults (4 pairs of legs)	Same as Tetranychidae	Egg – larvae – adult	Egg--Protonymph (2 pairs of legs) -- Deuteronymph (2 pairs of legs) –Adult (2 pairs of legs)
Red spider mite on okra, cotton, citrus, Jowar mite	Citrus flat mite	Yellow mite on chilli, Paddy panicle mite	Jasmine mite

Morphology of Mites

Body is vermiform, divided into cephalothorax and abdomen in family Eriophyidae. It contains 2 pairs of legs. In Tetranychidae, body is divided into Gnathosoma and Idiosoma. Gnathosoma contains mouth parts like Chelicerae and Pedipalpi that cover the mouth cavity. Above the mouth cavity there is a capitulum or tectum dorsally. Gnathosoma consists of 3 segments where the second segment has chelicerae and 3rd segment has pedipalpi. The Idiosoma is further divided into Podosoma and Opisthosoma. Podosoma has legs that are further divided into Propodosoma and Metapodosoma. Propodosoma has 2 pairs of legs and Metapodosoma has 2 pairs of legs. Opisthosoma is the posterior part of the body having anal opening. Eyes may be present or absent. In some mites if eyes are absent body surface act as photo sensitive organ. Mouthparts are Chelicerae, 3 segmented, modified into stylet like piercing organs. Pedipalpi are present on dorsoventral surface of Gnathosoma resembling the legs. These are modified as piercing or grasping organs. Legs may be 2 or 4 pairs. Each leg consists of coxa, trochanter, femur, genu, tibia and tarsus.

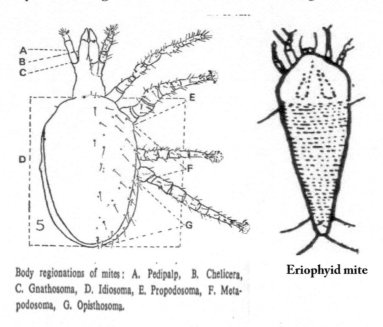

Body regionations of mites: A. Pedipalp, B. Chelicera, C. Gnathosoma, D. Idiosoma, E. Propodosoma, F. Metapodosoma, G. Opisthosoma.

Eriophyid mite

Habit and Habitat

Since most of the plant mites are negatively phototrophic, these are normally found on the under surface of the leaves either in colonies (like members of the families: Tetranychidae, Tenuipalpidae, Eriophyidae) or in solitary forms (like members of the family Tarsonemidae and majority of the predatory mites). But sometimes when the population on the lower surface increases considerably, some of the mites may also occur on the upper surface of the leaves. Some of the spider-mites cover their colonies with thin webs where all the developmental stages may be seen. Those who do not live in colonies are found either on leaf lamina or in angles formed by the major veins. A good number of mites (Phytoseiidae, Tydeidae etc.) may be found on twigs or stems or under barks. The eriophyid mites, the most peculiar among the mites, are often found on various malformed or deformed plant parts. A large number of mites may be found in the flowers, buds, axils of leaves, etc. The occurrence of mites within the galls is also not uncommon, but in case of mite galls, they are usually of open types contrary to the insect galls which are all of closed types. Some mites usually have very specialized habitats and are only found in the particular place in which they are adapted to live. For example, an animal pest mite will not live on a plant. Some mites, such as the two-spotted spider mite (*Tetranychus urticae*), are able to live on many different types of plants whereas others, such as the blueberry bud mite (*Acalitus vaccinii*), can only live in a certain part of a particular plant. Mites can be distinguished from their insect relatives by the presence of two body regions (cephalothorax and abdomen, in some these two are fused), four pairs of legs (only two pairs in Eriophyidae), sucking mouth parts and lack of antennae and wings. Mites possess chelicerae as mouth parts which are needle like useful for sucking sap from plants. Adults vary in body shape and possess 2 or 4 pairs of legs. The life cycle consists of an egg, larva, proto nymph, deuto nymph, trito nymph and adult stages. Oval shaped eggs are laid on leaves. Incubation Period is 6-13 days. The no.of nymphal instars vary among the families. Example: Eriophyid mite has only 2 nymphal instars and Tetranychid mite has 3 nymphal instars. The nymphs are active and nymphal period vary from 1- 3 weeks. The total life cycle in summer extend from 3-6 weeks.

Members of the order Acari are diverse, abundant, and sometimes economically important, though we know relatively little about them. Some plant-dwelling mites are beneficial by preying on other arthropods (Predatory), others cause no visible injury to plants, but some are serious plant pests (Phytophagous). Some notable species cause damage as parasites on different plants and animals. Species in this group have vastly different behavior. Some spend their lives beneath the soil or in the leaf litter, decomposing dead leaves or bits of organic matter (Saprophytic and detrivores). Others live in the nests of animals or upon an animal's skin. Many mites are ectoparasites of other organisms, while some prey on other arthropods. Still, others feed on plants or decomposed organic matter like leaf litter. There are even gall-making mites. Some are vectors of bacteria or other disease-causing organisms, making them a significant public health concern. These tiny organisms often disperse by attaching or riding on other more mobile animals. Such transport behavior is known as phoresy. Many mites have developed non predacious feeding habits (feeding on bacteria, yeasts, fungi, algae, mosses and higher plants. Most species of mites are predatory and will feed on a variety of small invertebrates, while others are more herbivorous and often feed on plant sap, sometimes causing damage to agricultural crops and garden plants.

Sampling and Identification of Mite Pests

Sampling of mites: Sampling involves collecting repeated systematic data of an organism in its environment over a specified time. Sampling unit is area within the sampling universe from which measurements are taken eg. traps, known numbers of plants or parts of plants, known numbers of sweeps with a sweep net. The objectives of sampling or monitoring are to detect the presence or absence of pests; quantify abundance of pests and their natural enemies; and follow the progress of an arthropod population through time by regular, periodic sampling.

Sampling of phytophagous mites

Direct counting: This is the commonest of all the methods. The sampled leaves are put into a petridish containing chloroform-soaked cotton to immobilize mites. Later, the leaves are put one by one under a stereo-binocular microscope and the mites are counted either from the entire leaf or from a definite leaf area depending on the intensity of population and intra leaf variation, if any. Comparing the plotted values based on actual counting of mites with the number of leaves having no mites, one can have a rough estimate of mite population.

Imprinting method: A whatman filter paper no. 1 is put over the under surface (because mites normally inhabit under surface) of a leaf and a wooden roller is rolled rapidly from base to apex of the leaf and back with pressure sufficient enough to crush all the mites. The mites after being crushed leave characteristic stains on the filter paper which makes permanent records of the population. Later, those spots are counted. An experienced person can easily differentiate the stains left by the adults and nymphs which are normally brownish and those left by the eggs which are paler.

Flotation method: The leaves are plucked from the fields and are put in glass jar containing water and a detergent (Teepol) and are then vigorously shaken. The mites are dislodged and then are collected by filtration. The residue is washed with alcohol in a cavity block to get the mites.

Jarring method: A white enamel tray with a cotton pad at the inner surface is kept under portion of the plant wherefrom the mites need to be collected. The plant part is beaten with a wooden rod which will cause dislodgment of mites and subsequently those will fall on cotton pad and get entangled. Later, those are picked up with a brush. Small mites as tydeids, etc. which are difficult to collect by heat treatment through funnel, can be easily collected by this method.

Brushing method: The leaf to be brushed is held between the brushes. The mites get dislodged and fall on a plastic disc kept below the brushes. Since the plastic disc is kept smeared with Vaseline, all those mites which fall on it get attached in vaseline. Later, these are picked up with a brush. The leaves should be worked forward and backward several times to ensure thorough dislodgment of all the mites from it. However, for counting of population from the sectors of plastic discs, there are definite formulae available and those are to be applied depending upon the population density of mite to be sampled.

Sampling of predatory mites: Some predatory mites inhabit on twigs and crevices of stems and therefore sampling of leaves may not give correct picture of its population. In such cases, samples should be made from all possible plant parts. In case of insect predators, some difficulty may arise because they may fly away or may fall off when the leaves or twigs are disturbed. The following methods may be used:

Counting after knocking down of predators: In this case, the predators are dislodged on a petridish after spraying or dusting a pesticide and those are counted.

Suction method: The predator can be collected quite easily by a suitable aspirator and then counted.

Visual search: Without disturbing the predators, a visual search over a certain number of plants or trees, if made, for a desired period of time, a reasonably correct picture about the population of predator can be have.

Trapping method: In this method, the predators are trapped by corrugated bands and then the population is counted.

Question-Answer

MCQs

1. The natural residence of every organism is known as?

a. Biome   b. Niche   c. Habitd   d. Habitat

Ans- d. Habitat

2. What is the name of the feature that allows organism to survive in the condition of its habitat?

a. Adjustment   b. Adaption   c. Acclimation   d. Adaptive Variation

Ans- b. Adaption

3. Mites belong to the class-

a. Insecta   b. Arachnida   c. Crustacea   d. Chilopoda

Ans: B. Arachnida

4. Wheat streak mosaic disease is transmitted by-

a. *Phytoseilus persimilis*   b. *Aceria cajani*   c. *Aceria tulipae*   d. *Acarus siro*

Ans: c. *Aceria tulipae*

5. Example of gall mite is-

a. Citrus flat mite   b. Paddy panicle mite   c. Jowar mite   d. Jasmine mite

Ans: d. Jasmine mite

6. The family of spider mite is

a. Tetranychidae   b. Brassicaceae   c. Eriophyidae   d. None

Ans: a. Tetranychidae

7. The mites that causes allergic reactions are

a. Varraoa mites   b. Nest mites   c. Dust mites   d. Itch mites

Ans: c. Dust mites

8. Mites are ____ in colour

a. Reddish-brown   b. Green   c. Blue   d. All of these

Ans: d. All of these

9. How many legs mites have.... 6/8/10/12

Ans: 8

10. Mites that commonly affects humans, those don't live in.... carpet/ grass/ furniture/ mattress

Ans: Grass

11. Pitfall traps are more useful in.... dry areas/ flooded areas/ both dry and flooded areas/ normal areas

Ans: Dry areas.

12. Sampling insects from soil is often.... complex and expensive/ simple but expensive/ simple and pocket-friendly/ complex still pocket friendly

Ans: Complex and Expensive.

13. Common household mites that feed on flakes of dead skin from humans and pets, are.... Itch mites/ bird mites/ Varroa mites/ dust mites

Ans: Dust mites.

14. How many antennae mites have? 2/3/1/0

Ans: 0

15. The largest mites measure about.... 5mm/ 6mm/ 7mm/ 10mm long

Ans: 6mm

16. The family of spider mite is:- a) Tetranychidae,b) Brassicaceae, c) Eriophydae, d) None

Ans. Option-"a"

Fill in the blanks

1. In species distribution ------------------ is spatially arranged.

Ans: Biological taxon

2. Distribution pattern may change by ---------, -------------, --------------.

Ans: Season, Humans, Availability of resources

3. Climatic factors consist of -----------, -----------, -------------.

Ans: Sunlight, Atmosphere, Humidity

4. Most of the mites are less than ------- mm in length.

Ans: 1

5. -------------- developed and used a special sampling work of brown wheat mites.

Ans: Henderson

6. Mites are preserved in _____

Ans: Ethyl alcohol

7. Mouthparts of mites are called _____

Ans: Chelicerae

8. Mites belong to the class _____

Ans: Arachnida

9. Mites are more closely related to _____ and _____

Ans:- Spiders and Ticks

10. Red palm mites are currently found in _____

Ans: South Florida

11. Mites that often feed on plant sap are _____

Ans: Herbivorous

12. Mites thrive by forming _____ relationship with other organisms

Ans: parasitic

13. In heating method, to remove mites from the leaves, ___ may be used

Ans: special brushing machine

14. In heating method, the plant sample harbouring insects may be placed in a special device, __ that heats the sample.

Ans: Berlese funnel

15. __ and __ mites can bury themselves under human skin and tissue

Ans: Scabies, chigger

16. Mites belong to the subclass _____

Ans: Acari.

17. Mites are _____ legged arthropods.

Ans: Eight.

True/ False

1. Humans are one of the largest distributors due to current trends.

Ans: True

2. Predation and disease are the biotic factors for species distribution.

Ans: True

3. Most of the mites have segmented body.

Ans: False

4. Most of the species of mites are harmless to human.

Ans: True

5. Pitfall trap is a useful sampling method for mites.

Ans: True 2.

6. In mites, respiration takes place only through skin.

Ans: False
(In general, they breathe by means of tracheae, or air tubes, but in many species, respiration takes place directly through the skin.)

7. The mite lifecycle consists of only three stages.

Ans: False. (The mite life cycle consists of four stages: egg, larva, nymph and adult)

8. Mite's body typically measure less than millimeter in length.

Ans: True

9. Mites are not ubiquitous.

Ans: False

10. Mites may spread wherever host animals travel.

Ans: True

11. Mites always show signs of socialization.

Ans: False

12. Researchers estimate that all species of mites are finally discovered.

Ans: False

13. Most species of mites are predatory.

Ans: True

14. Mites live on every single continent on Earth except Antarctica.

Ans: False

15. Pitfall trap involves direct observation and rely on good eye.

Ans: False

16. In heating method the temperature of insects' habitat is increased for forcing them to come out of their dwellings.

Ans: True

17. In Pitfall trap method, the trap collects both mature and immature insects which fall accidentally into it.

Ans: True

18. Mites legs are jointed.

Ans: True

19. Mites are bigger than ticks.

Ans: False

20. All mites are visible to the naked eye.

Ans: False

21. The smallest mites are about 0.5mm long.

Ans: False

22. Mites belong to Mollusca phylum.

Ans: False.

23. Mite does not affect on human.

Ans: False.

Short Questions and Answers

1. What is the scientific name of Citrus rust mites?

Phyllocoptruta oleivora

2. What is the host plants of Jowar mites?

Banana, sugarcane and sorghum.

3. What is the suitable habitat for Dust mites?

Dust mites can live in the bedding, mattresses, upholstered furniture, carpets or curtains in your home. Dust mites are nearly everywhere.

4. What is the morphological character of Red Vegetable Mites?

Red vegetable mite`s nymphs are yellow to brown after hatching & turn green, with dark specks along its side and adults are red to purplish red in colour.

5. What is the reproduction system of mites?

Reproduction in mites is very variable with some species mating through the direct transfer of sperm via coupling of the genital regions. Other species transfer sperm indirectly with the male placing a sperm droplet on the genital opening of the female with his legs or chelicerae

6. What is Focality?

The concept, known as "focality", provides a framework for prediction of sites of high density of mite population and allergen exposure, as well as a basis for manipulating the microenvironment for control purposes.

7. Which mite shows necrotic patches on green coconut?

Aceria guerreronis

8. What is the symptom of sweet potato rust mite?

The Leaves of sweet potato turn pale brown and cause severe rusting.

9. What is the habitat of False Spider Mites?

These mites severely damage the crops like pomegranate, guava, coconut and pointed gourd.

10. What is the body structure of Eriophyidae Mites?

Body is vermiform, divided into cephalothorax and abdomen in family Eriophyidae. It contains 2 pairs of legs.

11. Mention sampling techniques for predator mites.

Suction method, Trapping method, Visual search, Knocking down of predators.

12. What are the parts of leg of the mites?

Trochanter, Femur, Genu, Tibia, Tarsus

13. What are mites?

Mites are small arachnids. Mites are not a defined taxon, but the name is used for members of several groups of the class Arachnida.

14. What is sampling?

Sampling is a process used in statistical analysis in which a predetermined number of observations are taken from a larger population.

15. How can we identify a mite?

Similar in appearance to ticks but much smaller, mites have bulbous, round, or pill-shaped bodies. Classified as arachnids, mites have eight jointed legs. Their size varies by species, but most mites are usually invisible to the naked eye.

16. Write down the sampling methods of mites (only mention the method names)

In situ count, heating method, pitfall trap, extraction from the soil

17. Mention the taxonomic subclass of mite?

Acari

18. Write down some types of mites?

Phytophagous mites, predatory mites, dust mites, clover mites, bird mites, itch mites, varroa mites etc.

19. In situ count method is useful for which type of insects?

In situ count method is useful for counting large and conspicuous insects.

20. Which are the equipments used for smaller insects in In-situ count method?

The equipments that are used for smaller insect in in-situ count methods are hand lens, 1cm² windows that may be cut out in a thick paper sheet and placed on a leaf surface.

21. Write down the disadvantages of heating method?

In this method, sometimes insects may die during the extraction process. Thus, insects may not come out and not be counted.

22. What is gnathosoma?

Mites' bodies have a head like feeding appendage called gnathosoma

23. Why soil sampling is useful for insects population?

About 90% of insect species spend at least one stage in the soil or at the soil surface. Therefore, soil sampling can give useful information on the insect's population.

24. How sampling for soil pests may be done?

Sampling for soil pests may be done by digging in a quadrate or drilling the soil with some equipment and insect counts are made from a fix volume of soil.

25. Write down common techniques used for extraction of insects from soil sample?

Common techniques used for extraction of insects from soil samples are using: berlese funnels, sieving, washing and flotation. Soil sample may be taken with cores, golf hole bores, bulb cutter, shovels, trowels, metal frames or other equipments

26. Which mite can produce web?

Spider mites

27. Which one is 'false spider mite'?

Brevipalpus spp.

28. Name one chilli mite.

Polyphagotarsonemus latus.

29. *Aceria jasmine* mite affects which plant or crop?

Jasmine.

30. *Tetranychus neocalidonicus* (Red vegetable mite) causes damage on which crops?

Okra, green beans, mango, watermelon, *Chrysanthemum* and *Hibiscus*.

31. What are the key identification points of Citrus rust mite?

1. Adult body elongate and wedge-shaped, three times long than wide.

2. Extremely small.

3. Light yellow to straw coloured.

Chapter - 4

Survey on Pests and Forecasting of Pest Incidence

Introduction

Management of several of the important insect pests of crops is becoming increasingly difficult owing to a number of factors by which pests have been abled to offset the effect of control interventions. Several of the pest management problems may be ascribed to the excessive and indiscriminate use of chemical pesticides during last three decades. Routine application of insecticides for controlling insect pests without taking consideration about whether economic damage to the crops is being inflicted by pests or not has resulted into several problems like pest resistance, pest resurgence, pest replacement, killing of natural enemies of pests, presence of toxic residues in food and soil and environmental pollution etc. apart from increase in the cost of plant protection. Entomologists have been trying to find solution of these problems through Integrated Pest Management of which monitoring of insect pests activity through regular surveys is an important component. In fact knowledge of insect activity is indispensable since pest management system cannot operate without estimates of pests and natural enemies population without reliable assessment of plant damage and its effect on yield. Different activities of insects such as their population in the crop, damage inflicted to the plants, insect stage present, local movement, migration and dispersal etc. are documented through surveillance. The decision whether the control measures are needed or not depends upon accurate estimation of number. Further based on study of interrelations of pests population with various environmental factors, we should be able to predict future population trends or outbreaks of pests so that appropriate control measures are initiated when required. For example, when chemical control is to be adopted we must be able to forecast the need of spray or to predict most effective spray date in the light of likelihood of pest population reaching economic threshold level.

Survey: Survey is a planned activity to collect some data. Survey on pests is conducted to study the abundance of a pest species. It is detailed collection of insect population information at a particular time in a given area.

Types of survey – Qualitative survey, Quantitative survey, Roving survey and Fixed Plot survey.

» **Qualitative survey: (detection of pest)**

- Aimed at pest detection

- Provides list of pest species present along with reference to density like common, abundant, rare.

- Employed with newly introduced pests to understand the extent of infestation.

- Adopted at international borders (avoid invasion of any new species).

» **Quantitative survey: (enumeration of pest)**

- Define numerically the abundance of pest population in time and space.

- Provides information on damaging potential of a species and data can be used to predict future population trends.

- Provide the basis to decision making for adopting control measures.

» **Roving survey:**

- Assessment of pest population/damage from randomly selected spots representing larger area

- Large area surveyed in short period

- Provides information on pest level over large area.

- Survey is done from south to west direction by diagonal walk.

» **Fixed Plot survey:**

- Assessment of pest population/damage from a fixed plot selected in a field.

- The data on pest population/damage recorded periodic from sowing till harvest.

- e.g. 1 sq.m. Plots randomly selected from 5 spots in one acre of crop area in case of rice.

From each plot 10 plants selected at random.

Total tillers and tillers affected by stem borer in these 10 plants counted. Total leaves and number affected by leaf folder are observed. Damage is expressed as per cent damaged tillers or leaves. Population of BPH from all hills of 10 plants are observed and expressed as number/hill.

Insect pests	Methods
Adult stage of Jassids, Hispa, Gundhi bug and Stem borer	20 sweeps in the fixed field. Count the total no. of each insect
Jassid (nymphs), BPH (nymph & adult)	Tap vigorously 5 hills in 1 m² and count the insects fallen down on water and ground. Workout the population /hill
Stem borer, Rice gall midge, Leaf folder, Case worm	Compute the percentage of affected tillers and leaf from 5 hills/m²

Pest surveillance: Surveillance is the monitoring of the behavior, activities, or other changing information, usually of people, insects, and pathogens. It most usually refers to observation of individuals or groups by government organizations. For example, pest surveillance is the monitoring the progress of an insect/many insects in an agro ecosystem.

Pest Surveillance is the constant watch on population dynamics of pest, its incidence and¬ damage on each crop at fixed intervals, to fore-warn farmers to take-up timely crop protection measures. Regular monitoring of the pest will aid in decision making of pest management practices, and this can be achieved through pest surveillance. Pest surveillance can be done using the light traps, pheromone traps, food traps, attractants, pitfall traps (for soil insects), field scouting etc.

Importance and advantages of pest surveillance

» Useful for pest forecasting.

» Help to plan cropping pattern.

» Help to plan pest management programmes.

» Aids in developing models, and to find-out the Thumb Rule Models.

» Help in application of insecticides (stage, dose, type etc.).

» Helps in maintaining stability of Agro ecosystem.

Objectives of pest surveillance:

» To know existing and new pest species

» To assess pest population and damage at different growth stage of crop

» To study the influence of weather parameters on pest

» To study changing pest status (Minor to major)

» To assess natural enemies and their influence on pests

» To assess the effect of new cropping pattern and varieties on pest

Components of pest surveillance

» Identification of the Pest

» Distribution pattern and prevalence of the Pest

» Levels and severity of Pest

» Losses due to pest incidence

» Population dynamics

» Weather parameters.

» Data on natural enemies

Insect monitoring:

> » Insect Monitor means to maintain regular surveillance on insects aspects such as population, biology, movement and among many others.

> » Monitoring is the process by which the numbers and life stages of pest organisms present in a location are established.

> » The population size and the level of activity of beneficial organisms must also be determined for arthropod management.

> » There is no single monitoring technique that works for all categories of pests, and even within a pest category the best technique differ due to pest biology and ecology.

> » For example: direct counting of nematodes in soil would require the use of different techniques than counting tobacco hornworm caterpillars on a tomato plant

Why monitor? Often it may be difficult to identify the insects causing the particular damage. If this is the case, the next option available is to examine the symptoms of crop damage. We use the diagnostic tool to help to identify the insect pest causing the damage.

Importance of monitoring

> » To assess the pest situation and determine what sort of pest activity is occurring

> » For decision making

> » To predict pest problems before they occur

Monitoring techniques and procured

a. What to look for

i.   If unrecognized pests are found, samples should be collected and brought to the county extension office.

ii. Evidence of potentially contributing activities to the pest problem

iii. Presence of natural enemies

iv. Presence and evidence of pests

v. Evidence of damage

-Are there still pests present in the damaged area

-Where the damage is found

-Nature of damage

b. Determined by the biology of the pest

Determined by the crop, if a crop has a low threshold of damage, more intensive monitoring may be needed at regular intervals (weekly and may be more frequently when a pest approaches a borderline to becoming a threat to a crop)

c. Size of area to monitor

The field is surveyed in a pre-arranged pattern, such as walking in an S, U, Z, V or X shape, depends on the degree of accuracy required and the resources available, enough to provide good field representation or coverage, depends on the crop, the farm size, and the pest.

d. Record keeping

Accurate records are important for decision making and for evaluating trends in pest populations season to season. If a monitoring form is developed, it should provide information on

» Units of sample, e.g. insects per tree, infected fruits per plant

» Sample method

» Identification of the field and the sample date

» Both harmful and beneficial insects

Pest Surveillance and Monitoring in India

» Pest surveillance and monitoring form an integral part of IPM technology.

» Directorate of Plant Protection, Quarantine and Storage (DPPQS), Faridabad, is organizing regular rapid roving pest surveys on major field crops in different agro ecosystems in collaboration with ICAR and SAU's and a consolidated report then issued by Plant Protection Adviser (PPA) to the Government of India.

Every year there is huge loss to the agricultural yield and productivity due to pest and diseases, the reason being:

1. No proper and centralized database for analysis and forecasting of pest disease.

2. No on time advisory and early warning to the farmers to take appropriate action against the pest/ disease.

3. No authenticity of the survey data

4. Compartmentalized information and many more.

Types of reports involved in the surveillance programme are:

» **White card report:** this is normal report in which the pest studies are reported regularly at weekly interval.

» **Yellow card report:** is a special reporting system whenever pest is noticed at 50% of the economic threshold level but still not obtained EIL status.

» **Red card report:** the red card reporting system is adopted when pest has reached the critical ETL where immediate action has to be launched for controlling the pest.

Types of surveillance

a. **General surveillance:** Process whereby information on particular pest which is of concern for an area is gathered from many sources, wherever it is available and provided for use by NPPOs (National Plant Protection Organizations)

b. **Specific survey:** Procedures by which NPPOs obtain information on pest of concern on specific sites in an area over a defined period of time.

Pest Forecasting

Forecasting is the process of making statements about events whose actual outcomes (typically) have not yet been observed. Pest Forecasting is the systematic monitoring of pest population, dispersion and dynamics in different crop growth phases using models prepared based on the previous data, to forewarn the farmers to take-up timely crop protection measures needed. Advance knowledge of probable pest infestation (out breaks) in a crop would be very useful not only to plan the cropping pattern (to minimize the pest damage) but also to get the best advantage of pest management programs. Pest Forecast will help and guide us with insect timing and biology to eliminate blanket applications of pesticide, reduce pesticide amounts, and achieve quality results. Online pest forecasting data is developed by both public and private sectors. Forecasting of pest incidence often requires systematically recorded specific field data in an elaborate manner over considerable period of time which can be easily retrieved and analyzed.

Utility of Insect Forecasting

>> To predict the forth-coming infestation level of the pest, which knowledge is essential in justifying use of control measures mainly insecticidal applications.

>> To find-out the critical stage at which the applications of insecticides would afford maximum protection.

>> To assess the level of population and damage by pest during different growth stages of crops

>> To study the influence of weather and seasonal parameters on pest.

>> To fix-up hot-spots, endemic and epidemic areas of the pest.

>> To fore-warn farmers to make decisions in timing of control measures.

Types of Forecasting:

Short term forecasting: Covers a particular season or one or two successive seasons only, and is usually done by using the insect traps, often such predictions can also be based on rate of emergence of the pest (observed in insectory).

Long-term forecasting: Covers larger areas and is mainly dependent on the possible effects of weather on the insect abundance or by extrapolating the present population density. For long term forecasting more basic understanding of population dynamics of pest in relation to weather parameters is required.

Forecasting must be related to ETL: Effective monitoring and using economic thresholds make up the core for any IPM program. Control decisions are to be taken based on pest or damage monitoring data, potential damage, cost of control methods, value of production, impact of other pests, beneficial organisms and the environment. Decide if the cost of the pest management programme / pesticide application is justified in terms of the value of the commodity protected. When making decisions, consider the Economic Injury Level (EIL) and of Economic Threshold Level (ETL) where available.

EIL: The EIL is the pest population level when the loss caused by the pest is equal to the cost of control measures.

ETL: The ETL is the pest population or damage level when control measures are applied to keep the pest population from reaching the EIL.

Both the ETL and EIL values can change from year to year depending on the crop value and control costs.

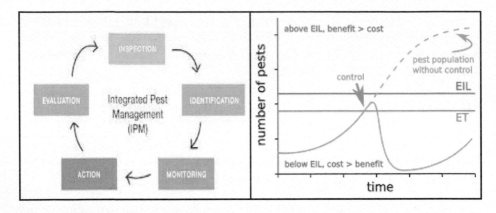

How to make the pest forecasting

Forecasting must be made through:

A. Population studies: Studies carried out over several years using appropriate sampling methods to find out seasonal range of population variability and geographic distributions. Using appropriate sampling techniques, the pest abundance must be studied over several years along with seasonal range, variability in number and distribution. The seasonal counts in relation to climate and topography need to be provided.

B. Studies on the pests life history: The possible no. of generations and behaviour of different larval instars, length of life cycle both in the field, overwintering, host range, number of eggs laid, etc. and other parameters can be studied in laboratory. Laboratory can be related to the range of environmental factors. Ex. Temperature, humidity.

C. Field studies of the effects of climate on the pest and its environments: Climatic factors not only affect the pest abundance but also affect the natural enemy population which is an important natural factor in controlling pest population. In field situations, the natural enemy abundance under a range of temperature and humidity should be studied. The other cultural practices like fertilizer application, irrigation, plant spacing, etc., affect the crop phenology which directly influences the population build-up of a pest.

D. Number of pest, parasites and predators: Life-table studies of pest are important for better understanding of pest population build-up, natural mortality factors, intrinsic growth rate, etc. Life-table of a pest can be helpful in finding mating and emergence period which are quite useful for predicting population dynamics of the pest. The migration and immigration of pests can also be used for forecasting of pests

Criteria for successful insect pest forecasting system

Cost effectiveness – Forecasting system should be cost affordable relative to available insect pest management tactics.

Multipurpose applicability – Monitoring and decision-making tools for several diseases and pests should be available.

Usefulness – The forecasting model should be applied when the insect can be detected reliably.

Importance - The insect pest is of economic importance to the crop.

Simplicity – The simpler the system, the more likely it will be applied and used.

Reliability – Use of sound biological and environmental data

Forecast Model - Types

» **Between year models** – These models are developed using previous years' data. The forecast for pests and diseases can be obtained by substituting the current year data into a model developed upon the previous years.

» **Within year models** – Sometimes, past data are not available but the pests status at different points of time during the current crop season are available. In such situations, within years growth model can be used, provided there are 10-12 data points between time of first appearance of pests and maximum or most damaging stage.

» **Weather related forecasting model:**

Observations-

  » Crop data: Phenological development, Growth, Leaf area and Variety

  » Insect pest: Pest population

  » Weather data required (hourly for ten or more years) – Precipitation, Temperature, Sunshine/cloudiness, Relative humidity, Leaf wetness, Wind direction and speed.

» **Degree-Day models:**

One (1) degree-day (DD)-DD is way of measuring of insect growth and development in response to daily temperature

» Degree-days (DD) are used in models because they allow a simple way of predicting development of cold-blooded organisms (insects, mites, bacteria, fungi, plants).

» Degree-day models have long been used as part of decision support systems to help growers to predict spray timing or when to begin pest scouting.

» Limitations of Degree-Day Models

• Insect response to temperature is not linear.

• Lower Thresholds Temperature known for very few species.

• Measured temperatures not the same as those experienced by the pest

» **Pest simulation models:**

Pre –requisites for Simulation models

• Mathematical descriptions of biological data.

• Computer programs or software to run these models.

• Application of these models in understanding population dynamics and dissemination of pest forecasts for timely pest management decisions.

Question – Answer

MCQs

1. Qualitative survey based on

   a. enumeration of pest   b. detection of pest
   c. damage by pests   d. natural enemy counting

   Ans: b. detection of pests

2. For soil insects sampling we use

   a. pitfall trap   b. water trap
   c. food trap   d. attractant

   Ans: a. pitfall trap

3. .... report card is used when pest population is noticed at 50% ETL but not in EIL.

    a. white   b. red

    c. yellow   d. blue

    Ans: c. yellow

4. For quick moving insects sampling is done by —

    a. shaking   b. knockdown

    c. narcotized collection   d. trapping

    Ans: c. narcotized collection

5. Pheromone traps are used mainly for —

    a. adult male   b. nymph

    c. larva   d. female insects

    Ans: a. adult male

6. When loss caused by pests is equal to cost of control measures

    a. GEP   b. ETL

    c. EIL   d. Pest damage

    Ans: c. EIL

7. Which of the following is/are a category of pest survey?

    a. Qualitative survey   b. Quantitative survey

    c. Fixed plot survey   d. All of the above.

    Ans. d) All of the above.

8. The first step of pest surveillance is

    a. Determination of pest population   b. Estimation of yield loss

    c. Identification of pest   d. Estimation of abundance of natural enemies.

    Ans. c. Identification of pest.

9. Which of the following is not required to generate basic information to develop forecasting models?

    a. Quantitative seasonal study   b. Life-history study

    c. Anatomical study   d Ecological study.

    Ans. c. Anatomical study.

10. Which of the following can be placed under pest forecasting types?

    a. Short term forecasting   b. long term forecasting

    c. Both b and c   d. None of the above.

    Ans. c. both b and c.

11. Long term population forecast based on Markov chain theory was developed for effective management strategies for

    a. *Nilaparvata lugens*,   b. *Sogatella furcifera*,

    c. none of the above   d. both a and b.

    Ans. d. both a and b

12. Blue sticky traps are used as pest monitoring tools for

    a. Thrips   b. Mantids

    c. Aphids   d. Beetles

    Ans. a. Thrips

13. The trap capture data serves several purposes

    a. Ecological studies.

    b. Tracking insects migration.

    c. Timing of pest arrivals into agro-ecosystems

    d. All of the above.

    Ans: d. All of the above.

14. The process of collecting insects from an area by removing from crop and counting calls:

    a. Knock down   b. In situ counts

    c. Netting   d. Trapping

    Ans. a. Knock down.

15. Pest forecasting has generally based on:

    a. Environmental factor   b. Climatic areas

    c. Empirical observation.   d. All of the above.

    Ans. d. All of the above

16. Emergence trap is used as pest monitoring tool for:

    a. Soil insects   b. Sucking insects

    c. Flying insects   d. Phototropic insects.

    Ans. a. Soil insects

17. Which of the following is not true for forecasting?

    a. Forecasts are rarely perfect

    b. The underlying casual system will remain same in the future

    c. Forecast for group of items is accurate than individual item

    d. Short range forecasts are less accurate than long range forecasts

    Ans: d

18. In pest surveillance ……....Involves the in-situ assessment of pests and diseases in the standing crop

    a. The fixed plot surveys.   b. Roving survey.

    c. Both a and b.   d. None of the above.

    Ans: c

19. Primary factors influencing ETL & EIL –

    a. Management cost   b. Degree of injury per pest

    c. Crop susceptibility to injury   d. Both b and c

    Ans: a

20. Snail meal trap is used for

    a. Rice hispa   b. Rice gundhi bug

    c. YSB   d. Rice skipper

    Ans: b

21. For sampling, in case of hoppers, which stages are counted

    a. Egg   b. Nymph

    c. Adult   d. Both b and c

    Ans: d

22. The temperature range comfortable for European corn borer (threshold temperature)-

    a. 50–85-degree F   b. 55–90-degree F

    c. 70-80-degree F   d. 60-90-degree F

    Ans: a

23. Which one provides information on pest level over large area?

    a. Fixed plot survey   b. Quantitative survey

    c. Roving survey   d. Qualitative survey

    Ans: c. Roving survey

24. Survey is conducted to study –

    a. The abundance of pest   b. Pest population

    c Crop growth   d. Diseases

    Ans: a. The abundance of pest

25. Which type of surveys are done on randomly selected spots?

    a. Qualitative   b. Roving

    c. Quantitative   d. Field type

    Ans: b. Roving

26. Fundamentals of sampling –

    a. Randomization and choice   b. Unit

    c. Pest count   d. None

    Ans: a. Randomization choice

27. Basic information about crop loss by pest is gathered by –

    a. Models   b. Implements

    c. Tools   d. Research

    Ans: d. Research

28. Forecasting mainly deals with –

    a. Crop production   b. Crop damage

    c. None   d. Surveillance

    Ans: c. None.

29. Which areas of pest incidence are marked –

    a. Endemic   b. Regular

    c. Both   d. None

    Ans: A. Endemic.

30. How many types of survey are there-

    a. 1   b. 2

    c. 3   d. 4

    Ans: b

31. How many types of forecasting are there –

    a. 1   b. 3

    c. 5   d. 2

    Ans: d

32. Which of these is not a components for Pest Surveillance-

    a. Identification of the Pest

    b. Distribution pattern and prevalence of the Pest

    c. Decrease natural enemies of pest

    d. Levels and severity of Pest

    Ans. c

Short Answer question

1. By which method nymph of jassids are counted?

5 hills of 1 sq metre are vigorously tapped and then fallen insects are counted.

2. When survey of same place is carried out at regular interval to record observation what does it call?

Surveillance.

3. How many types of report are in surveillance programme?

White, yellow and red card report.

4. How many types of pest population study are there?

Extensive and intensive study.

5. What is sample?

A group of sampling units is called sample from which an estimate is made.

6.   How severity of pest infestation can be measured?

By counting the number of eggs, larvae, pupae or flying insects, or by determining the percentage of infested leaf, if feasible.

7.   No. of organisms per unit area can be calculated by which formula?

Density (D)= Z/ 2R(V+W)^1/2. D= density Z= no. encountered in unit time. R = distance V= average speed of observer W= average speed of organism

8.   Define pest surveillance.

Pest surveillance refers to the constant watch on the population dynamics of pests, its incidence and damage on each crop at fixed intervals to forewarn the farmers to take up timely crop protection measures.

9.   Write the three basic components of pest surveillance.

The three basic components are --- Determination of a. the level of incidence of the pest species. b. the loss caused by the incidence. c. the economic benefits, the control will provide.

10. Write the two objectives of pest surveillance.

i) To estimate the crop losses caused by pests.

ii) To monitor when the pest population / damage at different growth stages of crop reaches the economic threshold levels.

11. Classify pest forecasting.

Pest forecasting can be classified as--- i) Short term forecasting ii)Long term forecasting.

12. What are the steps of pest surveillance?

i) Identification of pest.

ii) Determination of pest population.

iii) Estimation of abundance of natural enemies.

iv) Estimation of yield loss

13. In case if fixed plot survey, the survey is done over a (large area/ selected area)

Selected area

14. What do you mean by a sample?

A group of sampling units from which an estimation is made.

15. Which type of insects are caught by sweep net?

Grasshoppers, odonates.

16. Write the name of any two trapping method.

Light trap, pheromone trap.

17. What do you mean by survey?

Survey is a planned activity to collect some data.

18. Write the definition of pest forecasting.

This is the process of making prediction of the future based on past and present data and most commonly by analysis of trends.

19. On which type of crop forecast system is applicable?

Economic crop

20. What are the main data of a forecasting system?

Weather data, crop data and pathogen data.

21. What is the main objective of Pest forecasting?

To reduce yield loss

22. What is Pest Monitoring? or Describe Pest monitoring using one or two sentences ?

Monitoring for Pests is a fundamental first step in creating a proper integrated pest management (IPM) programme. Pests are monitored through a variety of monitoring tools such as pheromone traps, light traps, coloured sticky traps, pitfall traps and suction traps.

23. Mention the objectives of Pest Monitoring.

The objectives of Pest Monitoring are written down below. They are, (i) Estimation of changes in insect distribution and abundance. (ii) Information about insects, life history. (iii) Influence of biotic and abiotic factors on pest population.

24. Describe Pest surveillance using one or two sentences.

Pest surveillance refers to the constant watch on the population dynamics of pests, its incidents and damage on each crop at fixed intervals to forewarn the farmers to take up timely crop protection measures.

25. Write down the objectives of Pest Surveillance:

Three objectives of Pest Surveillance are written down below: a) To study the influence of weather parameters on pest by recording the changes in density of pest population throughout the year. b) To asses natural enemies population and their influence in a particular cropping system and in different reasons. c) To monitor the development of biotypes, resistance to insecticides, resurgence etc and also to estimate crop losses caused by pests.

26. Write down the fixed plot survey:

In fixed plot survey, the pest population on damage due to insect pests is assessed from a fixed plot selected in a field. The datas are recorded regularly from sowing till harvest of the crop from the same fixed plot in a particular field. The datas, collected in these surveys are used to develop forecasting models.

27. Write down the Rapid / Roving survey in few sentences.

This survey includes the assessment of pest population on damage from randomly selected spots in a short period of time over a large area. It provides information on pest level which helps in determining the timing of adopting appropriate control measures. The surveys are made to monitor the initial development of pests in endemic area in the beginning of crop season

28. Mention the methods for Pest forecasting.

The pest forecasting has generally based on the three methods- (i) Environmental factors. (ii) Observations of climatic areas – (a) zone of natural abundance (Endemic), (b) zone of occasional abundance, (c) zone of possible abundance (iii) Empirical observations.

29. What EIL stands for?

The full form of EIL is economic injury level. It is defined as the lowest population density that will cause economic damages (stern *et al.* 1959)- also defined as a critical density where the loss caused by the pest equals the cost of control measure. EIL can also be calculated using following formula where,

C/VIDK, where C = Cost of management activity per unit of production (Rs. /ha)

V = Market value per unit of yield or product (Rs. /tonne)

I = Crop injury per insect (Per cent defoliation/insect)

D = Damage or yield loss per unit of injury (Tonne loss/% defoliation)

K = Proportionate reduction in injury from pesticide use

30. What ETL stands for?

The full form of ETL is economic threshold level or also known as (Action threshold). ETL is defined as the pest density at which control measures should be applied to prevent an increasing pest population from reaching economic injury level (EIL). It represents pest density lower than EIL to allow time for initiation of control measures.

31. What are the techniques of sampling?

(i) Absolute sampling – To count all the pests occurring in a plot. (ii) Relative sampling – To measure pest in terms of some values which can be compared over time and space. Eg. Light trap catch & pheromone trap.

32. What is pest forecasting?

Forecasting of pest incidence or outbreak is based on information obtained from the pest surveillance.

33. What are the uses of pest forecasting?

The uses of pest forecasting are written down below: (i) Predicting pest outbreak which needs control measures. (ii) Suitable stage at which control measures give maximum protection.

34. Write down the importance of pest surveys.

It provides information on the damaging potential of a species and data can be used to predict future population trends.

35. Mention any two variety of traps used in pest monitoring.

The two variety of traps used in pest monitoring are – (i) Sex pheromone traps. (ii) Light traps.

36. What is pest phenology?

The growth of an insect to certain discrete stages, such as the larval, pupal or adult stages, is often referred to as development. When the development is linked with temperature, it is referred to as phenology.

37. What are the applications of remote sensing in pest management?

Basically the applications of remote sensing in pest management are noted below: (i) Photography and videography from aircraft and from the ground. (ii) Satellite – borne multispectral scanning (MSS) (iii) Thermal imaging (iv) Ground based and air borne radar methods (Riley, 1989)

38. What are the materials, tools and equipments used in conducting pest monitoring in the farm?

The common 'tools of the trade' include:

(i) Flashlight (ii) Blacklight (detect rodent urine) (iii) Video-camera (iv) Screw driver (v) Putty knife (vi) Spatula (vii) Tracking patches or powders. (viii) Double-sided transparent tape (tree & shrub insects).

39. How many types are there in pest surveillance?

There are three types of surveillance, mentioned below: (i) Fixed plot survey (ii) Random survey (iii) Roving survey

40. Write down few sentences about Qualitative survey on pest incidence.

Qualitative survey: It is generally aimed at pest detection and provides list of pest species present along with references to density like common, abundant, and rare. These are usually employed with newly introduced pest to understand the extent of infestation also adopted at international borders where agricultural commodities are inspected to avoid invasion of any new species.

41. Write down few sentences about Quantitative survey on pest incidence.

This survey defines numerally the abundance of pest population in time and space. It provides information on the damaging potential of a species and data can be used to predict future population trends. These surveys provide the basis to decision making for adopting control measures for a pest by the farmers.

42. Which survey is used to detect and enumerate pest?

The survey that is used to detect and enumerate pest is qualitative survey (detection of pest).

43. What is pest scouting?

Pest scouting is a process to determine no of insect pests and their crop loss in a specific crop field.

44. Difference of Survey and Surveillance.

Survey is making a single observation and recording pest population and whereabouts, whereas, Surveillance is repeated standardised survey so that changes can be detected.

45. How pest population is determined?

Three methods that help in determining the pest population are – absolute estimates, relative estimates, population indices.

46. What are Pheromone traps?

Synthetic sex pheromones are placed in traps to attract males are called pheromone traps. The rubberized septa, containing the pheromone lure are kept in traps designed especially for this purpose and used in insect monitoring / mass trapping programmes.

47. How pest incidence is calculated?

The intensity of incidence or pest infestation of a disease is usually measured by taking ratio of infected plants to the total number of plants in a selected plot of the field at various stages of the growth period.

48. What are the types of Pest- forecasting? Write down few sentences describing the types.

The two different types of Pest- forecasting are – A. Short term forecasting based on 1 or 2 seasons B. Long term forecasting based on 1 or 2 seasons.

49. Importance of Pest survey.

It provides information in the damaging potential of a species and data can be used to predict future population trends.

50. What do you know about the CAPS program?

This programs is a joint venture taken by state agriculture Department and USDS in Texas to look over initial stages of pest incidence and crop losses.

51. Usually short term forecasting is done by based on what?

Usually short term forecasting is done by based on the rate of emergence of the pest (observed in insectory).

52. How insect forecasting service is useful for farmers?

Applying pest and disease control is important to protect the farm and crops from the insects. Weather forecast helps the farmers to know when to apply the pests and chemicals to avoid the crop wastage. By some estimates, up to 40 percent of the world's food supply is already lost due to pests.

53. What is forecasting based on?

Forecasts are based on opinions, intuition, guesses, as well as on facts, figures, and other relevant data. All of the factors that go into creating a forecast reflect to some extent what happened with the business in the past and what is considered likely to occur in the future.

54. What is the full form of APHIS?

The Animal and Plant Health Inspection Service (APHIS). It is an agency of the United States Department of Agriculture (USDA) based in Riverdale, Maryland responsible for protecting animal health, animal welfare, and plant health.

55. Why monitoring is important in pest management decisions?

Monitoring pest populations is important to determine the onset of the pest in the crop being protected, seasonal patterns of the pest population, movement of the pest in the crop, life stage of the pest, abundance and distribution of the pest in the crop, and finally to assess the effectiveness of management tactics.

State True or False

1. When pest has reached the critical ETL yellow card is adopted.

Ans: False.

2. Line transact methods are used for grasshoppers.

Ans: True.

3. Aerial nets are used for soil borne insects.

Ans: False

4. Water trap are used for rice pests.

Ans: True.

5. In degree day model insect growth and development is based on daily temperature.

Ans: True.

6. The ETL is the pest population or damage level when control measures are applied to keep the pest population from reaching the EIL is called ETL.

Ans: True.

7. Short term forecasting is based on 1 or 2 seasons

Ans: True.

8. Quantitative survey is useful for detection of pest

Ans: False

9. When pest level crosses ETL, control measures have to be taken to prevent pest from reducing EIL

Ans: True.

10. Bait traps are used as pest sampling units for sorghum shoofly.

Ans. True.

11. Larger sample size does not give accurate results.

Ans. False.

12. Qualitative survey is used for detection of pest

Ans: True

13. ETL (economic threshold level) is defined as the lowest population density that will cause economic damage

Ans: False

14. EPIPRE is a system of supervised control of diseases and pests in rice

Ans: False

15. Yellow card report is adopted when pest has reached the critical ETL

Ans: False

16. In extensive programme, of a large area is done

Ans: True

17. Pheromone traps are species specific

Ans: True

18. Short term forecasting is based on 1 or 2 seasons.

Ans: True

19. Quantitative survey is useful for selection of pest

Ans: False

20. When pest level crosses ETL, control measures have to be taken to prevent pest from reducing ETL

Ans: True

21. Pest Forecasting is the systematic monitoring of pest population.

Ans: True

22. Short-term forecasting covers larger areas.

Ans.: False

23. In fixed plot survey a large area is surveyed in short period.

Ans: False

24. Survey and surveillance is same – F

25. Long term forecasting depends on weather – T

26. Damage of pests tend to vary through the life of individual plant – T

27. Parasitoid and predators are quantified through birth rates – F

28. The data for conserved gene sequences is generated under iBOL project – T

29. In Roving survey large area surveyed in short period. - T

30. Long term forecasting is done by based on rate of rate of emergence of the pest.
Ans: False

31. One objective of pest survey is to study changing pest status(Minor to major).
Ans: True

Fill in the blanks

1. Pest surveillance, monitoring form an integral part of_____ technology.
Ans: IPM (Integrated Pest Management)

2. _____ is organizing regular rapid roving pest surveys on major field crops in different agro ecosystems in collaboration with ICAR and SAU's
Ans: Directorate of Plant Protection, Quarantine and Storage (DPPQS), Faridabad

3. In white card report pest studies are reported at _____interval.
Ans: weekly

4. A proportion of the habitat from which insect counts are to be made is called_____
Ans: sampling unit.

5. The habitat in which population occurs is _____
Ans: sampling universe

6. To monitor when pest population/ damage at different growth stages of crop reaches the economic threshold levels, is an objective of ....
Ans: pest surveillance.

7. To define numerally the abundance of pest population in time and space, ... survey should be applied.
Ans: quantitative

8. The survey of pest population or damage from randomly selected spots in a short period of time over a large area can be termed as ....
Ans: roving survey.

9. The prediction of a particular pest depends upon .... of a pest and meteorological factors.

Ans: biology

10. The farmers can be aware of the timing and biology of insect incidence through .... of pests.

Ans: forecasting

11. Yellow sticky traps are used to attract a broad spectrum of .... insects.

Ans: flying

12. Pheromone traps are generally used to attract .... of specific pests such as the *Heliothis* moth.

Ans: males

13. Most pest forecast models take into account the .... of the herbivore and its host.

Ans: phenology

14. Light traps have been widely used for monitoring the population dynamics of lepidoptera & ....

Ans: coleoptera.

15. To count all the pests occurring in a plot, ... sampling technique is being followed.

Ans: absolute

16. Headquarters of Directorate of plant protection quarantine & storage (DPPQ&S) of India is located at ....

Ans: Faridabad.

17. Two types of pest forecasting are .... and. ....

Ans: Short term, Long term forecasting

18. Quantitative survey defines numerically the .... in time and space.

Ans: abundance of pest population

19. Pest forecasting is done to find-out the critical stage at which the applications of insecticides would afford _____ protection.

Ans: maximum

20. Survey is conducted to study the _____

Ans: abundance of the pest species.

21. Prediction of a particular pest depends upon ……….. of the pest and meteorological factors.

Ans: biology

22. CAPS programme is to survey pests that are …………………

Ans: exotic

23. …………..surveys are adopted to assess crop loss.

Ans: Direct loss

24. Sampling unit should be ………… and should not overlap.

Ans: distinct

25. The data collected in field plot survey is used to develop …………… models.

Ans: forecasting

Chapter - 5

Pest Surveillance Through Light Traps, Pheromone Traps and Forecasting of Pest Incidence

Pest Surveillance through Light Traps

Certain pests required positioning of various kinds of traps. Light trap is one of them to monitor the initial pest build up. Therefore, the State Department of Agriculture is to initiate action for positioning of these kinds of traps at strategic locations at village level.

Light traps, with or without ultraviolet light, attract certain insects. Light sources may include fluorescent lamps, mercury-vapor lamps, black lights, or light-emitting diodes. Designs differ according to the behaviour of the insects being targeted. Light traps are widely used to survey nocturnal moths. Total species richness and abundance of trapped moths may be influenced by several factors such as night temperature, humidity and lamp type. Grasshoppers and some beetles are attracted to lights at a long range but are repelled by it at short range.

Working principle

Light traps have been a proven method of pest control for decades. The UV rays emitted from the light trap lamps attract flies and other pests toward the trap. Then, depending on whether the trap uses high-voltage electricity or sticky glue-boards, the flies are both zapped, potentially causing insects to explode and spraying their body parts several feet from the zapper, or they are trapped in a layer of glue. Fly light traps that utilize electricity will include a pan at the bottom of the trap to catch the dead pests.

Using Lights to Attract Insects

i. A great number of insect species are attracted to light of various wavelength. Although different species respond uniquely to specific portions of the visible and no visible spectrum (as perceived by humans), most traps or other devices that rely on light to attract insects use fluorescent bulbs or bulbs that emit ultraviolet wavelengths (black lights).

ii. Hundreds of species of moths, beetles, flies, and other insects, most of which are not pests, are attracted to artificial light. They may fly to lights throughout the night or only during certain hours. Key pests that are attracted to light include the European corn borer, codling moth, cabbage looper, many cutworms and armyworms, diamondback moth, sod webworm moths, peach twig borer, several leaf roller moths, potato leafhopper, bark beetles, carpet beetles, house fly, stable fly, and several mosquitos. Hundreds of species of moths, beetles, flies, and other insects, most of which are not pests, are attracted to artificial light.

iii. A light trap used to survey night-flying insects. Most light traps use ultraviolet lamps and capture a wide range of moths, beetles, and other insects.

iv. The UV light trap can be placed in food grain storage godowns at 1.5 m above ground level, preferably in places around warehouse corners, as it has been observed that the insect tends to move towards these places during the evening hours. The trap can be operated during the night hours. The light trap attracts stored product insects of paddy like lesser grain borer, *Rhyzopertha dominica*, red flour beetle, *Tribolium castaneum* and saw toothed beetle, *Oryzaephilus surnamensis* in large numbers. Psocids which are of great nuisance in godowns are also attracted in large numbers. Normally 2 numbers of UV light trap per 60 x 20 m (L x B) godown with 5 m height is suggested. The trap is ideal for use in godowns meant for long term storage of grains, whenever infested stocks arrive in godowns and during post fumigation periods to trap the resistant strains and left over insects to prevent build up of the pest populations.

Different Types of Light Traps

Incandescent light trap: They produce radiation by heating a tungsten filament. The spectrum of lamp include a small amount of ultraviolet, considerably visible especially rich in yellow and red. Example: Simple incandescent light trap, portable incandescent electric trap. Place a pan of kerosenated water below the light source.

Mercury vapour lamp light trap: They produce primarily ultraviolet, blue and green radiation with little red (e.g.) Robinson trap. This trap is the basic model designed by Robinson in 1952. This is currently used towards a wide range of Noctuids and other nocturnal flying insects. A mercury lamp (125 W) is fixed at the top of a funnel shaped (or) trapezoid galvanized iron cone terminating in a collection jar containing dichlorvos soaked in cotton as insecticide to kill the insect.

Black light trap: Black light is a popular name for ultraviolet radiant energy with the range of wavelengths from 320-380 nm. Some commercial type like Pest-O-Flash, Keet-O-Flash are available in market. Flying insects are usually attracted and when they come in contact with electric grids, they become electrocuted and killed.

Pest Surveillance through Pheromone Traps:

Insect pheromones

Pheromones are chemicals released into environment in small amounts by special abdominal glands in insects. Pheromones are species specific, may stimulate one gender or all genders. Male moths detect pheromones with antennae. Synthetic sex pheromones are manufactured & used as lures.

Pheromone traps

Pheromones are chemicals for species- specific communication. Most often, these sex pheromones are produced by females to attract males and are most well known for adult lepidoptera. Commercially produced by synthesizing and blending the appropriate chemicals, the sex pheromones are loaded into dispensers, which can be placed in traps of various designs for deployment in agriculture, horticulture, forestry and storage. A pheromone trap is a type of insect trap that uses pheromones to lure insects. Sex pheromones and aggregating pheromones are the most common types used. A pheromone-impregnated lure, as the red rubber septa in the picture, is encased in a conventional trap such as a bottle trap, delta trap, water pan trap, or funnel trap.

Pheromone traps are used both to count insect populations by sampling, and to trap pests such as clothes moths to destroy them. Synthetic sex pheromones are placed in traps to attract males. The rubberized septa, containing the pheromone lure are kept in traps designed especially for this purpose and used in insect monitoring / mass trapping programmes. Water pan trap and funnel type models are available for use in pheromone based insect control programmes.

Traps baited with synthetic sex pheromones are useful in estimating population and detecting early stages of pests. Pheromone traps however attract only target pest even when crop is attracted by many other pests.

Application of pheromone traps:

Synthetic analogues of sex pheromones of quite large number of pests are now available for use in pest management. Sex pheromones are being used in pest management in different ways.

Monitoring:

Monitoring is an effective way to determine the population trends of insects and plays a critical role in pest management programmes. It involves the use of a synthetically derived pheromone formulated into a dispenser and trap to selectively attract and intercept the target insect. Pheromones can be used for monitoring pest incidence/outbreak in the following ways

» To detect the introduction of a pest in a new area

» To assess whether the pest has assumed economically serious level or there any need for control measure

» To time the control measure.

Attraction and killing (Mass trapping): In this approach insects are attracted to a source and killed by one means or another. Techniques in attract and kill programs range from entanglement in sticky materials to outright killing with insecticides or pathogenic micro-organisms. By mass trapping large number of one sex of insects are killed and hence mating success is reduced, and the population in next generation falls. The pheromone trap used in mass trapping is same as those used in sampling and detection but the trap densities vary. For mass trapping, 100 traps per hectare are used compared to 5 to 10 per hectare in sampling.

Mating disruption by air permeation (Confusion or Decoy method): In this method synthetic pheromone is permeated into the environment to mask the natural pheromone and thus disrupt the normal pheromonal communication among insects. Such disruption will cause failure of insects to locate their mates thereby prevent mating. Formulations like flakes, hollow fibres and microcapsules containing pheromone are applied to the field to cause mating disruption.

Different pheromone traps

a) Funnel trap b) Plastic moth trap   c) Delta trap   d) Nomate trap

Advantages of pheromone traps:

• Are affordable •If used wisely, can detect low insect populations •Insect identification not needed • Easy to install & manage • Nontoxic, no residue on food. • Can be used season long.

Disadvantages of pheromone traps:

• May be cumbersome to handle • Take precautions when handlings (gloves). Several traps could difficult to manage • Knowledge intensive – may need tech support • Weather sensitive. • Does not tell about plant injury.

Different types of pheromone lures

S. No.	Type with particulars	Target pest	Name of Lure
1.	Pheromone trap for brinjal fruit and shoot borer (Wota-T®): It is easy to assemble on a single pole consisting of an adaper, basis to hold water (mixed with kerosene or detergent) and a lure holder with a canopy	*Leucinodes orbonalis*	Lucin-lure
2.	Pheromone trap for paddy stem borer (Fero-T®): Consists of a funnel base with handle for tying to a stick, a canopy to be fixed on 3 pegs provided on the funnel and transparent sleeve.	*Scirpophaga incertulus*	Scirpo-lure
3.	Pheromone trap for tobacco catterpillar (Fero-T®)	*Spodoptera litura*	Spodo-lure
4.	Pheromone trap for american boll worm (Fero-T®)	*Helicoverpa armigera*	Heli-lure
5.	Pheromone trap for melon or oriental fruit fly (Fligh-T®) having yellow bowl with entry hole and translucent dome	*Bactrocera cucurbitae.*	Bacu-lure

Forecasting of pest incidence: Pest forecasting is the perception of future activity of biotic agents, which would adversely affect crop production. In other words, it is the prediction of severity of pest population which can cause economic damage to the crop. The systematically recorded data on pest population or damage over a long period of time along with other variable factors, which affect the development of pest, may be helpful in forecasting the pest incidence. The prediction of a particular pest depends upon characteristics/biology of a pest and the meteorological factors. These meteorological factors may affect the pest either directly affecting their survival, development, reproduction, emergence and behaviour, or indirectly by their action on host plants or on natural enemies. These factors also determine the geographical limits of distribution and the time of appearance and abundance of pests. The forecasting of pests guides the farmers about the timing and biology of insect incidence, and to eliminate blanket applications, reduce pesticide amounts, and achieve quality results. The farmers can take to timely action of applying various pest control measures to harvest maximum returns. In fact several studies are required to generate the basic information, which is required to develop forecasting models.

Forecasting must be related to ETL: Effective monitoring and using economic thresholds make up the core of any IPM program. Control decisions are to be taken based on pest or damage monitoring data, potential damage, cost of control methods, value of production, impact of other pests, beneficial organisms and the environment. Decide if the cost of the pest management programme / pesticide application is justified in terms of the value of the commodity protected. When making decisions, consider the Economic Injury Level (EIL) and of Economic Threshold Level (ETL) where available.

EIL: The EIL is the pest population level when the loss caused by the pest is equal to the cost of control measures.

ETL: The ETL is the pest population or damage level when control measures are applied to keep the pest population from reaching the EIL.

Both the ETL and EIL values can change from year to year depending on the crop value and control costs.

How to make the pest forecasting

Forecasting must be made through:

A. Population studies: Studies carried out over several years using appropriate sampling methods to find out seasonal range of population variability and geographic distributions. Using appropriate sampling techniques, the pest abundance must be studied over several years along with seasonal range, variability in number and distribution. The seasonal counts in relation to climate and topography need to be provided.

B. Studies on the pests life history: The possible number of generations and behaviour of different larval instars, length of life cycle both in the field, overwintering, host range, number of eggs laid, etc. and other parameters can be studied in laboratory. Laboratory can be related to the range of environmental factors. Ex. Temperature, humidity.

C. Field studies of the effects of climate on the pest and its environments: Climatic factors not only affect the pest abundance but also affect the natural enemy population which is an important natural factor in controlling pest population. In field situations, the natural enemy abundance under a range of temperature and humidity should be studied. The other cultural practices like fertilizer application, irrigation, plant spacing, etc., affect the crop phenology which directly influences the population build-up of a pest.

D. Number of pest, parasites and predators: Life-table studies of pest are important for better understanding of pest population build-up, natural mortality factors, intrinsic growth rate, etc. Life table of a pest can be helpful in finding mating and emergence period which are quite useful for predicting population dynamics of the pest. The migration and immigration of pests can also be used for forecasting of pests.

Different Types of Forecasting Study

Quantitative seasonal studies: Using appropriate sampling techniques, the pest abundance must be studied over several years along with seasonal range, variability in number and distribution. The seasonal counts in relation to climate and topography need to be provided.

Life-history studies: The detailed bio-ecology of pest under a range of temperature, humidity, etc. should be known. The duration of different instars, number of generations, survival rate, amount of food eaten, overwintering, host range, number of eggs laid, etc. and other parameters can be studied in laboratory.

Ecological studies: Life-table studies of pest are important for better understanding of pest population build-up, natural mortality factors, intrinsic growth rate, etc.

Life-table of a pest can be helpful in finding mating and emergence period which are quite useful for predicting population dynamics of the pest. The migration and immigration of pests can also be used for forecasting of pests.

Field studies: Climatic factors not only affect the pest abundance but also affect the natural enemy population which is an important natural factor in controlling pest population. In field situations, the natural enemy abundance under a range of temperature and humidity should be studied. The other cultural practices like fertilizer application, irrigation, plant spacing, etc., affect the crop phenology which directly influences the population buildup of a pest.

Types of Pest Forecasting: Pest forecasting may be divided into two categories, viz., short-term forecasting and long-term forecasting.

1. Short-term forecasting: The short-term forecasts are often based on current or recent past conditions that form a basis for, or an enhancement to, the forecast. These may cover a particular season or one or two successive seasons only. The pest population is sampled from a particular area within a crop using appropriate sampling technique and the relationship is established between weather data and progress in pest infestation. The laboratory studies on the effect of temperature on emergence and egg laying can be used to forecast the pest situation in the field. The short term forecasting can be completely empirical, such as use of environmental cues reported from Japan, where the date of first blooming of cherry blossom and the mean March temperature were used to predict the peak emergence of rice stem borer. Based on multiple regressions, short term forecasting of wheat grain aphid, *Sitobion avenae* (Fabricius), has been done. The peak population density on each field was positively correlated with the population densities at the end of ear emergence, mid-anthesis and the end of anthesis. Based on two counts on the crop, the accuracy increased from ear emergence to the end of anthesis, however, the forecast at mid-anthesis of peak density was much more accurate.

2. Long-term forecasting: These forecasts are based on possible effect of weather on the pest population and cover a large area. The data are recorded over a number of years on wide seasonal range and from different areas. Long-term forecasting is based on knowledge of the major aspects of the pest insect's lifecycle, and of how it is regulated. The data recorded are analyzed and models are developed based on the available information. The models help in forecasting pest population in various geographical areas based on common weather parameters. Long-term population forecast based on Markov chain theory was developed for effective management strategies for *Nilaparvata lugens* (Stal) and *Sogatella furcifera* (Horvath). This model is an effective method for long-term population forecasting of *N. lugens* and *S. furcifera*, and thus provides plant protection agencies and organizations with valuable information in implementing appropriate management strategies. Long-term forecasting of brown and white backed plant hoppers in Japan was based on the assumption that both the hoppers overwinter as diapausing eggs on winter grasses. After it was discovered that the brown plant hopper migrates in Japan from outside, the short-term forecasting was adopted.

Factors for Pest Forecasting

Pest forecasting has generally based on environmental factors, climatic areas and empirical observations.

1. Environmental factors: The population development of a particular pest mainly depends on the favourable environmental conditions available in a particular geographical region. The pest attack occurs in epidemic form only when the favourable environmental conditions for multiplication of pest prevail for longer duration. Therefore, the factors responsible for environmental conditions are the major criteria on the basis of which the forecasting can be done. The sugarcane pyrilla, *Pyrilla perpusilla* (Walker), outbreak is predicted based on high temperature during monsoon. The population per 30 plants (Y) is predicted based on the mean maximum temperature (X) of week preceding the data of observation of field population. Insects are exothermic (cold-blooded) and their body temperature and growth are affected by their surrounding temperature. Biological development of insects over time in correlation to accumulated degree days has been studied, discovering information on key physiological events, such as egg hatch, adult flight, etc. There is a threshold temperature for each insect; for example, 48°F for the alfalfa weevil, *Hypera postica* (Gyllenhal). No development occurs when temperatures are below that level. Insects have an optimum temperature range in which they will grow rapidly. Then, there is maximum temperature (termed upper cutoff) above which development stops. These values can be used in predicting insect activity and appearance of symptoms during the growing season. Therefore, the degree days would be useful in pest management programme to time the scouting of insect pests. This predictive information is known as an insect model. Models have been developed for a number of insect pests. Degree days = Maximum temperature + Minimum temperature/2 – Development threshold As an example, codling moth, *Cydia pomonella* (Linnaeus), pheromone monitoring traps are placed in the apple orchard at 100 degree days after March 1 in northern Utah to determine initiation of adult moth flight. A temperature range of 50° to 85°F is most comfortable for European corn borer, *Ostrinia nubilalis* (Hubner). Below 50°F, it will not develop, and above 85°F, development will slow dramatically. A degree day for European corn borer is one of degrees above 50°F over a 24- hour period. For example, if the average temperature for a 24-hour period was 70°F, then 20 degree days would have accumulated (70 − 50 = 20) on that day. These accumulations can be used to predict when corn borers will pupate, emerge as adults, lay eggs, and hatch as larvae.

2. Observations of climatic areas: The distribution of insects throughout the world is based on evolutionary history which includes main important factor, i.e., climate of the geographical region. There are three distinct zones of abundance of each insect species. Zone of Natural Abundance (Endemic): In this zone, the pest species is often in large number, regularly breeds and is a regular pest of some importance. The climate conditions are most favourable for its development and pest is seen all the time. Zone of Occasional Abundance: The insect species emerge in epidemic occasionally in this zone because the climatic conditions are either less suitable or the suitable conditions exist only for a short period of time followed by unsuitable conditions. Sometimes, the climatic is severe to destroy the entire population, which is then re-established by dispersal from zone of natural abundance. Zone of Possible Abundance: The pest species in this zone can be seen only after migration from zone of natural and occasional abundance outbreaks. The climatic conditions are drastic for their breeding and development. The population is destroyed by the severe climatic conditions within a short period of time. Three different regions Orlando, Naples and Ankara corresponding to zone of natural abundance, occasional abundance and possible abundance, respectively are known for Mediterranean fruit fly, *Ceratitis capitata* (Wiedemann). The observation on the climatic areas where critical infestations are likely to occur can be predicted for some insects. Combination of climatic factors like temperature, rainfall, humidity, etc. existing in a geographical region gives an indication of possibility of establishment of pest in that region. The other factors like biotic and topography may also be used for prediction of insect pests.

3. Empirical observations: This type of pest forecasting is based on estimating the number of insects available during a particular time. In other words, it is nothing but the sampling of insect or monitoring of pest population. It involves forecasting the population in the next season by counting the pest in the previous seasons. In many cases, the number of pests in the early part of cropping season will give an indication as to the extent of its likely multiplication in the season. From the counting of immature stages of insects, approximate estimations of later stages can be made. For example, in india taking soil cores for insect eggs of carrot fly, *Psila rosae* (Fabricius) and cabbage root fly, *Delia radicum* (Linnaeus), is successful for estimating the later population of root maggots. The adult catch in the traps especially pheromone traps can be used to estimate the approximate abundance of pest population later in the season. The sampling of insect pest on alternate host/ weeds during non-availability of main crop can be quite useful to forecast the pest population development in the coming season, e.g., counting overwintering eggs of blackbean aphid, *Aphis fabae* Scopoli, on spindle trees helps in estimating the aphid population on peach-potato crop. In many lepidopteran species, pest forecasting is based on estimating the number of eggs and young larvae on the crop, e.g., cotton bollworms, stem borers, pulse moths, etc

Some forecasting model:

The CIPRA: (Computer Centre for Agricultural Pest Forecasting) software allows the user to visualize forecasts of insect development. Within CIPRA; there is forecasting models for a total of 35 pests (25 insects and 10 diseases). In vegetables (cabbage, carrot, onion, potato, tomato), fruits (apple, grape, strawberry), and cereals (wheat, barley, corn). Each crop has its own independent computer programs file, known as DLL (Dynamic Link Library), which makes possible their integration in other specialized software.

The forecasting models database: It is user friendly software that can predict the development of pests, crops and some post-harvest disorders based on hourly weather data. In real time the software allows to target best time for pest control approaches. This software is based not only on the weather observations from several automatic stations but also on weather forecasts. The CIPRA approach contributes significantly to pesticide reductions in the environment and promotion of sustainable crop production systems. It uses mathematical models and hourly weather data and forecasts, to produce reports that facilitate the evaluation of infection or sporulation of some diseases, development of insect population.

Computer Software Model for Prediction for Insect Pest

> » PEST-MAN is a computerized forecasting tool for apple and pear pests- Canada.

> » MORPH is predictive computer model for horticultural pest- UK.

> » SOPRA is applied as a decision support system for eight major insect pests

of fruit orchards - Switzerland and southern Germany.

» The SIMLEP decision support system for Colorado potato beetle (*Leptinotarsa decemlineata*)- Germany and Austria.

» ECAMON to predict the presence or absence of the European corn borer in Czech republic during the 1961 -1990 (reference period).

» It helped to explain the sudden increase in maize infestation during the periods of 1991-2000.

» It also estimates that this potential niche will expand within the next 20-30 years.

Rice pest

» A model simulating yield loss due to several rice pests under a range of specific production situations in tropical Asia.

» It was developed by the IRRI in Phillippines.

Success of a forecasting: The success of a forecasting system depends, among other things, on the commonness of epidemics (or need to intervene)

» The accuracy of predictions of epidemic risk (based on weather, for example)

» The ability to deliver predictions in a timely fashion

» The ability to implement a control tactic (Insecticide application, for example)

» The economic impact of using a predictive system

Question-Answer

MCQs

1. What is pest surveillance?

a) Watching your pesticide application kill the pest
b) Record keeping of the pesticide used
c) Checking or scouting for pests in an area to determine what pests are present
d) The pest's predators.

Ans. c

2. Which is the basic component of pest surveillance?

a. Determination of the level of incidence of the pest species
b. Determination of what loss the incidence will cause.
c. Determination of economic benefits or other benefits the Control will provide
d. All of the above

Ans. d

3. What is/are the objective(s) of pest surveillance?

a. It is the way of detecting different species of harmful pests.

b. It is the way of detecting the change in behaviour of the pests.

c. IPM is a way in which natural enemy of different harmful insect pests are found.

d. All of the above.

Ans: d

4. Surveillance is often called as –

a. Survey   b. Incidence
c. Estimation   d. Monitoring

Ans: d

5. Light trapping of insects is done in the:

a. Cultural methods   b. Legal methods
c. Mechanical methods   d. Preventive methods

Ans. c

6.   Light trap art used for the collection of:

a. Positively phototropic   b. Negatively phototropic
c. Geotropic   d. Hydrotropic

Ans: a

7.   Most adult insects are attracted towards light in –

a. Night   b. Dawn
c. Noon   d. Daytime

Ans. a. Night

8.   Light traps are widely used to survey

a. Nocturnal moth   b. BPH
c. Mosquito   d. None

Ans: a

9.   In which colour the insect attracts most?

A. Black   B. Green
C. Red   D. White

Ans: A

10. Who discovered Mercury vapour lamp?

A. F Smith   B. Robinson
C. Main kyon   D. None

Ans : B

11. What range of wavelength is used in black light trap?

A. 320 – 380 nm   B. 500nm
C. 600nm   D. 800 nm

Ans : A

12. Which trap does attract especially male insects?

A. Pheromone trap   B. Light traps
C. Bait trap   D. Pit Fall Trap

Ans: A

13. Pheromone coated paper strips in the confusion technique are thrown over an area to

(a) Confuse males so that they are unable to locate females
(b) Repel insects from a region
(c) confuse females so that they are unable to locate males
(d) Attract insects and kill them

Ans: a

14. What are the tools of attractant traps?

a. Light trap   b. Pheromone trap
c. Sticky trap   d. All

Ans: all

15. What are the main functions of pheromone trap?

A. Monitoring   B. Attraction and killing
C. Mating disruption   D. All of the above

Ans : All of the above

16. Which is not a pheromone trap?

A. Funnebars   B. Nomate Trap
C. Delta trap   D. Net trap

Ans: D

17. Forecasting of pest is the

A. The prediction of severity of pest population which can cause economic damage to the crop.
B. The prediction of pest population density
C. Both A and B
D. None

Ans: A

18. How many types of pest forecasting exist?

A. 1   B. 2
C. 3   D. 4

Ans: 2

19. Which Factors affect on pest forecasting?

A. Environment   B. Climatic
C. Emperial   D. All of the above

Ans: D

20. Pheromone traps attract –

a. Pollinators   b. Dispersal agents
c. All pests present   d. Target pest only

Ans: d. Target pest only

21. Locust warning stations are an example of –

a. Short term monitoring
b. Long term monitoring
c. Long term forecasting
d. Short term forecasting
Ans: c. Long term forecasting

22. What is/are the objectives of Pest Surveillance?

(i) It is a way of detecting different species of harmful insects
(ii) It is a way of detecting the change in behaviour of the pest
(iii) IPM is a way in which enemy of different harmful insect-pests are found
(iv) All of the above

Ans: Option (iv)

23. Farrow Light trap can capture flying insects for its-

(i) Strong Light   (ii) Large base
(iii) Net like accessory   (iv)Funnel like

Ans: Option (ii)

24. Pheromone traps attract-

(i) Pollinators   (ii) Dispersal agents
(iii) All pests present   (iv) Target pest only

Ans: Option (iv)

25. Locust warning stations are an example of-

(i) Short term monitoring   (ii) Long term monitoring

(iii) Long term forecasting   (iv) Short term forecasting

Ans: Option (iii)

26. In pest surveillance ....involves the in-situ assessment of pests and diseases in the standing crop –

(a) The fixed plot survey   (b) Roving survey

(c) Both a and b.   (d) None of the above.

Ans: C. Both a and b.

27. Methyl eugenol trap is effective against.

(a) Butterflies   (b) Rice moth

(c) Fruit flies   (d) Hairy caterpillar.

Ans: c. Fruit Flies

SAQs

1. What is pest surveillance?

Checking or scouting for pests in an area to determine what pests are present is called pest surveillance.

2. What are basic components of pest surveillance?

a. Determination of the level of incidence of the pest species b. Determination of what loss the incidence will cause. c. Determination of economic benefits or other benefits the control will provide.

 3. Name some methods by which pest surveillance is done?

Light trap, pheromone trap

4. Why are insects attracted to UV light?

Flying insects are most likely attracted to ILTs for Warmth. For example, house flies prefer temperature about 82.4 degrees Fahrenheit, which is warmer than most interiors, so they seek warmer areas and are attracted to the ILT.

5. Name of some insect pests which are attracted by light trap

Nocturnal moths, BPH

6. Name different types of light trap

Mercury lamp light trap, black light trap etc.

7. What is insect pheromone?

Insect Pheromones are chemicals used by an insect to communicate with other members of the same species.

8. How do Pheromone Traps work?

They work in one form of mating disruption; males are attracted to a powder containing female attractant pheromones. The pheromones stick to the males' bodies, and when they fly off, the pheromones make them attractive to other males. It is hoped that if enough males chase other males instead of Females, egg laying will be severely impeded.

9. What are the application of pheromone trap?

Monitoring, attraction and mating disruption

10. What are the different types of pheromone trap?

Delta trap, Plastic moth trap

11. What are the advantages and limitations of pheromone trap?

 Advantages: easy to manage and Nontoxic

Disadvantages: Training required

12. What do you mean by a pheromone trap?

A pheromone trap is a type of insect trap that uses pheromones to lure insects. Sex pheromones and aggregating pheromones are the most common types used.

13. What is forecasting of pest incidence?

Pest forecasting is the perception of future activity of biotic agents, which would adversely affect crop production.

14. What is short term forecasting?

Short term forecasting: Covers one or two seasons mainly based on the populations of the pest within the crop by sampling methods.

15. What is long term forecasting?

Long-term forecasting is based on knowledge of the major aspects of the pest insect's life- cycle, and of how it is regulated.

16. Name of some forecasting model

CIPRA, ECAMON

17. How is forecasting made?

Forecasting is made through population studies carried over several years, studies on the pest life history, field studies on the effect of climate on the pest and its environment.

18. Name a few light sources used in light trap.

Light sources may include fluorescent lamps, mercury-vapour lamps, black lamps or light emitting diode.

19. How long does pheromone moth traps last?

In general all pheromone lures are good for 60 days.

20. Which light is useful for insect pest control for trapping?

The ordinary light trap consists of an electric bulb emitting yellow light as attractant and a funnel to direct lured into a container containing water.

21. Why should an average farmer use Pest Surveillance?

Pest surveillance is the corner stone for pest management through which epidemic situations can be avoided by detecting damage prior to establishment at a higher pest population.

22. What is the unit of one fixed plot?

The unit of a fixed plot is one acre (4000 sq.mtr).

23. How many sub-plots are there in a fixed plot?

There are 5 sub-plots in a fixed plot.

24. What are the pest monitoring devices generally used in e-pest surveillance?

Generally Pheromone trap, sweep net, square meter frame, water pan, sticky trap, sphere trap, sieves and light traps are use as Pest monitor in e-Pest surveillance.

25. What are the physical requirements for e-Pest surveillance?

Manpower, Transport (Mobility), Equipment, Smart Phone & facilities for data processing are the requirement for surveillance.

26. Besides pests what others are surveyed at the time of surveillance?

A. Natural enemies (Parasites, Predators). B. Location of the field. C. Water level in the field (Flooded, Muddy, Dry) D. Weathers (Sunny, Cloudy, Drizzling, Raining). E. Crop Variety (Local, Improved, Hybrid, Tolerant, and Resistant).

27. What are the uses of pest forecasting?

Predicting pest outbreak which needs control measure and suitable stage at which control measure gives maximum protection.

Fill in the blanks

1. ------------- is the constant watch on population dynamics of pest, its incidence and damage on each crop at fixed intervals.

Ans: Pest surveillance

2. ----------------are widely used to survey nocturnal moths.

Ans. Light traps

3. Grasshoppers and some beetles are attracted to lights at a ------------ range but are repelled by it at ----- range.

Ans: Long, short

4. Locust warning station in India was established in _____.

Answer: 1939.

5. Pheromones are secreted by ------ gland

Ans: Abdominal

6. Pheromone traps are effective against _____.

Ans. Spotted boll worm, Dimond back moth, White grubs.

7._____ is used to predict the forth-coming infestation level of the pest.

Ans. Pest Forecasting

8. Pest Surveillance is done to know the .... of certain pest in a field.

Ans. Population dynamics

9. ..... is a planned activity which is conducted out at regular intervals to get precise data.

Ans. Surveillance

10. The short-term forecasts are based on _____.

Answer - current or recent past conditions

11. The best part of light trap data is to _____ but for the tricky part is _____.

Ans. I. show the presence and relative abundance of particular species, II. quantitative analysis

12. Short term forecasting is done through ....

Ans. Sampling method

True or False

1. Sex pheromones are used in both pheromone traps and light traps.

Ans. False

2. Grasshoppers are attracted to lights at a short range.

Ans. False

3. Survey is conducted occasionally while surveillance is conducted regularly.

Ans. True

4. A nationwide pest forecasting system was established in Japan in 1942.

Ans. False

5. Pest surveillance can not assess natural enemies and their influence on pests.

Ans. False.

6. Light traps work best in forests. (True/False)

Ans. False

7. UV light is the most efficient as light trap. (True/False)

Ans. True

8. Pheromone traps leaves residue on the crop which causes damage to the crop.

Ans. False

9. Pheromone traps can be widely used because they are more affordable than other options.

Ans. True

10. Pest forecasting helps to fix-up hot-spots, endemic and epidemic areas of the pest.

Ans. True

11. Short-term forecasting is based on knowledge of the major aspects of the pest insect's life- cycle, and of how it is regulated.

Ans. False

12. CIPRA is a model of pest forecasting

Ans. True

13. Pest surveillance does not maintain the stability of agro-ecosystem.

Ans. False

Chapter - 6

Identification of Pests
and Their Estimation

Objectives of Pest Identification

To know the pest and its damage in crop plants which is the prerequisite for the effective pest management.

Materials: The materials required for the lab and field diagnosis is depend on the type of crop plants and its parts to be observed or diagnosed. Some of the common materials required are sharp knife, magnifying lens, scissor, sample container, microscope, camel hair brush, spade etc.

Diagnosis of pest damage: Insect and non-insect pests cause a particular type of damage to the plant parts often characteristic to particular pests. The pest mostly insect not present on the site of damage makes it difficult to know the causal agent. Sometimes, the symptoms of damage caused by insects may closely resemble to those resulted due to pathogens or due to nutritional disorders. So, practical experience makes familiar about the correct diagnosis on the basis of visual symptoms of damage to take appropriate control measures.

It is very important to know and study about the symptoms of damage by various phytophagous insect pests for proper, effective and economic management. Insect-pests are found to cause injury to plant either directly or indirectly to secure food, their development and further generations. Insect-pests attack on various parts of plant viz., root, stem, bark, leaf, bud, flower and fruit. Based on the nature and symptoms of damage, insect-pests can be classified into different groups.

Moreover, the process of diagnosing a plant health problem without any specialized laboratory equipment is called field diagnosis. This is the situation at a plant clinic and when making a farm visit. Many times farmer brought the plant sample to us, they may have collected the wrong part of the plant or the sample may have deteriorated in transit. It may be necessary to visit the field to see fresh symptoms and to gain other information on the pest. If you intend to send a sample to a colleague or a formal diagnostic support service, it is usually a good idea to visit the field yourself and to select a fresh sample of your own.

General points to take in consideration during diagnosis

» Don't be in too much of a hurry. Slow down, cut open the plant and have a look inside.

» Use a hand lens to look for small insects and their stages.

» Most importantly, talk to the farmer and ask questions relating to what you are looking at and the ideas that are forming in your mind

The following is a summary of what to do when visiting a field to observe the symptoms of the entire crop:

Step 1: Get in close	• What parts are affected? • Describe symptoms using the correct terminology. • Observe changes in shape, colour and growth. • Look for visible signs of insect pests and non-insect pests
Step 2: Look at the whole plant (including Roots)	• Where are the symptoms within the plant? • Which growth stages are affected? • How do the symptoms progress from early to late stages? • How severe is the attack?
Step 3: Examine groups of plants	• Incidence: how many plants are affected? • Distribution: random, edge of the plot only, in patches, pattern caused by use of machinery? • Remember: consider plant variety, age and how it is grown.
Step 4: Speak To farmers and other local extension workers	• When did the problem appear? Is this the first time? • Record local name for the problem. • Consider soil type and climate (patterns) • Obtain information on the varieties used, recent history of chemical inputs used, etc.

Damage to leaves

No of Symptoms	Insect group and crop	Example
1. Leaf margin notched	Ash weevils on cotton and brinjal	*Myllocerus spp.*
2. Leaf Lamina scarified	Thrips on cotton	*Thrips tabaci*
3. Young terminal leaves drying in nursery	Rice thrips	*Stenchaetothrips biformis*
4. Leaves silvered and wilting	Onion thrips	*Thrips tabaci*
5. Young terminal leaves curling upwards along margin	Chillies thrips Gall thrips	*Scirtothrips dorsalis* *Liothrips karny*
6. Leaf edges curled with honey dew, sooty mould and ant movement	Cotton aphid, Jassid on cotton and bhendi, Grape vine mealy bug, Coconut scale, Cotton whitefly, Curry leaf psyllid	*Aphis gossypii* *Amrasca devastans* *Maconellicoccus hirsutus* *Aspidiotus destructor* *Bemisia tabaci* *Diaphorina citri*
7. Leaf rolled longitudinally	Rice leaf roller Cotton leaf roller	*Cnaphalocrocis medinalis,* *Sylepta derogata*
8. Many small shot holes	Radish flea beetle	*Phyllotreta downsei*
9. Silt like small cuts like T shaped slits	Grapevine flea beetle	*Sceledonta strigicollis*
10. Larger regular shaped holes	Tortoise beetle (Sweet potato)	*Aspidomorpha milaris*
11. Irregular shaped holes	Many lepidopteran caterpillars and grasshopper, Polyphagous caterpillar	*Spodoptera spp.*
12. White parallel streaks along long axis	Hispa or spiny beetle	*Dicladispa armigera*
13. White parallel streaks along cut leaf tip forming leaf tube	Rice case worm	*Paraponyx stagnalis*
14. Graminaceous leaf cut laterally and rolled longitudinally	Rice skipper	*Pelopidas mathias*
15. Leaflets rolled	Coconut skipper	*Gangara thyrsis*

16. Leaf mine broad with central black faceal pellete and leaf edge folded dorsally for pupation, broad, streak like serpentine and papery mine	Citrus leaf miner	*Phyllocnistis citrella*
17. Lamina with ladder like windowing leaving veins intact	Epilachna beetle on brinjal and bitter gourd	*Epilachna spp*
18. Longitudinal margin blotching	Rice whorl maggot	*Hydrellia sasakii*
19. Grazing like cutting of seedlings	Rice swarming caterpillar or cut worm	*Spodoptera spp*
20. Leaves webbed together	Mango shoot webber, Sapota leaf webber	*Orthaga exvinacea Nephopteryx eugraphella*
21. Leaves fastened together to form nest	Red tree ant	*Oecophylla smaragdina*
22. Leaf cut laterally and rolled across	Mango, sapota leaf twisting weevils	*Apoderus tranquebaricus*
23. Parallel or serial holes	Sorghum stem borer	*Chilo partellus*
24. Hopper bum	Rice BPH, Cotton jasids	*Nilaparvata lugens, Amrasca devastans*
25. Leaf skeletonized with papery appearance	Early instars of cutworm on cotton, castor, banana, Cabbage diamondback moth	*Spodoptera litura Plutella xylostella*
26. Semi circular leaf cut	Leaf cutting bee on rose, guava and red gram	*Megachile anthracina*
27. Leaf folded and mined	Groundnut and redgram leaf miner	*Aproaerema modicella*
28. Round and elongate galls	Mango leaf galls (gall midges) Pungam leaf galls	*Amradiplosis spp. Eriophyid mite*
29. Eranium growth (felt like spongy hairy growth)	Jasmine eriophyid mite	*Aceria jasmini*
30. Minute yellow specks on leaf	Banana tingid, Coconut tingid	*Aspidiotus destructor, Stephanitis typicus*
31. Bunchy top	Banana	*Pentalonia nigronervosa f. typica*

32. Bubble froth or spittlemass on leaf or leaf axil	Spittle bugs on graminaceous plants	*Cercopidae sp.*
33. Hanging leaf cases	Case worm or bag worm	*Nymphula depuntalis*
34. Little leaf	Brinjal	*Cestius phycitis*
35. Phyllody	Gingelly	*Orocious albinictus*
36. Complete defoliation	Hairy caterpillars (Castor) Sphingid hawk moth (Gingelly)	*Euproctis fraterna, Acherontia styx*
37. Front with **V** shaped cut	Coconut rhinocerous beetle	*Oryctes rhinoceros*

Damaged flower buds and shoots:

Symptoms	Insects groups and crops	Example
1. Flower petals and perianth destroyed	Blister beetle on redgram and cotton	*Mylabris pustulata*
2. Flower petals with small holes	Cotton flower weevil	*Amorphoidea arcuata*
3. Flower buds bored	Moringa bud worm, Sapota budworm, Jasmine budworm	*Noorda moringae, Anarsia epotias, Hendecasis duplifascialis*
4. Squares damaged	Cotton spotted bollworm, Cotton spiny bollworm	*Earias vitella E. insulana*
5. Rosetted flowers	Cotton pink bollworm	*Pectinophora gossypiella*
6. Interlocular damage	Cotton pink bollworm	*P. gossypiella*
7. Silk damaged	Maize earworm	*H. armigera*
8. Capitulum damaged	Sunflower capitulum borer	*H. armigera*
9. Aborted flower	Moringa midge	*Stictodiplosis moringae*
10. Withering and shedding of flowerbuds and flowers	Mango hopper	*Idioscopus Spp.*
11. Inflorescence webbed	Mango flower webber	*Eublemma versicolor*
12. Blighted inflorescence	Cashew mirid bug	*Helopeltis antonii*

Damaged fruits and seeds:

Symptoms	Insects groups and crops	Example
1. Capsule damage	Castor capsule borer	*Dichocrocis punctiferalis*
2. Fruit bored	Brinjal fruit borer Mango fruit borer	*Leucinodes orbonalis* *Bactrocera dorsalis*
3. Boll damage	Cotton bollworm	*Helicoverpa armigera*
4. Ear heads with chaffy grains	Rice earhead bug, Sorghum earhead bug, Stink bug	*Leptocorisa acuta, Calocons angustatus, Nezara viridula*
5. Ear heads with chaffy grains and protruding pupal cases	Sorghum gall midge	*Contarinia sorghicola*
6. Webbing of grains in the earhead	Sorghum webworm	*Cryptoblabes gnidiella* *Antoba silicula*
7. Cob damaged	Maize earworm	*H. armigera*
8. Pod bored	Pulse pod borers, Gram pod borer, Plume moth, Spotted pod borer,Spiny pod borer	*H. armigera, Exelastis atomosa,Maruca testulalis, Etiella zinckenella*
9. Pod shrivelled with shrivelled grain inside	Pulse pod bugs	*Riptortus pedestris* *Clavigralla gibbose*
10. Necrotic spots on fruits and pods	Mirid bugs on guava fruit and cocoa pod	*Helopeltis antonii*
11. Flowers and young capsules with galls	Gingelly gall midge	*Asphondylia ricini*
12. Coconut scarred	Eriophyid mites	*Aceria guerreronis*
13. Citrus fruit with necrotic lesions,rotting and dropping of fruit	Citrus fruit sucking moth	*Othreis spp.*
14. Fruit and berry surface corky	Grapevine berry thrips,Banana fruit thrips, Cardamom thrips	*Scirtothrips dorsalis, Chaetanaphothrips Spp. Scirtothrips cardamomi*
15. Berries damaged	Pepper pollu beetle	*Longitarsus nigripennis*
16. Holes on stored cereal grains (rice, sorghum, maize, wheat etc.)	Rice weevil, Lesser grain borer	*Sitophilus oryzae, Rhyzopertha dominica*
17. Cereal grains with exit hole and flap door	Angoumois grain moth (Paddy, sorghum, maize, cumbu)	*Sitotroga cerealella*

18. Pin head size holes on processed tobacco	Cigarette beetle	*Lasioderma serricorne*
19. Pin head size holes on spices	Drug store Beetle (turmeric, coriander, ginger)	*Stegobium paniceum*
20. Pulse seeds with circular holes and white egg cemented on surface	Pulse beetle (All pulse grains)	*Callosobruchus maculatus*
21. Ground floor infested	Red flour beetle	*Tribolium castaneum*
22. Dense webbing of cereal grains	Rice moth (Rice, maize, sorghum etc.)	*Corcyra cephalonica*

Damaged stem:

Symptoms	Insect group and crop	Example
1. Dead heart	In graminaceous crops, Stem borer, shoot fly (sorghum, cumbu)	*Scirpophaga incertulas, Chilo spp, Atherigona spp*
2. White ear	Rice stem borer	*Scirpophaga incertulas*
3. Silver shoot or onion leaf	Rice gall midge	*Orseolia oryzae*
4. Extra tillering	Sorghum shoot fly	*Atherigona spp*
5. Terminal shoot with bunchy appearance	Sugarcane top shoot borer	*Scirpophaga excerptalis*
6. Apical part of the stem bored and drying	Spiny boll worm in cotton,	*Earias spp*
7. Tree trunk and branch with eaten bark, frass and silk tube	Bark borer in sapota, mango etc.	*Indarbela spp*
8. Tree trunk or branchwith round emergence hole and withering of branches	Stem borer (mango and other trees)	*Batocera rufomaculata*
9. Galled distal shoot	Stem gall fly (Cucurbits) Shoot weevil (Cotton)	*Neoclasioptera falcata Alcidodes affaber*
10. Stem galls	Stem weevil (Cotton) Stem weevil (Amaranthus)	*Pempherulus affinis, Hypolixus truncatulus*
11. Pseudostem bored and oozing of fluid	Banana pseudo stem borer	*Odoiporus longicollis*

12. Stem part encrusted and drying with scale	Sugarcane scale, Cassava scale	*Melanaspis glomerata* *Aonidomytilus albus*

Damaged root and tubers:

Symptoms	Insect group and crop	Example
1. Rhizome extensively bored	Banana rhizome weevil	*Cosmopolites sordidus*
2. Tube damaged	Potato white grubs, Sweet potato weevil	*Holotrichia spp., Cylas formicarius*
3. Roots with extensive swelling	Root knot nematode (Pulses)	*Meloidogyne Spp.*
4. Roots stunted and bushy	Lesion nematode (Citrus)	*Tylenculus spp.*
5. Roots with small round cyst	Cyst nematode (Potato)	*Heterodera spp.*

Damage to sown seed and seedling:

Symptoms	Insect group and crop	Example
1. Seedlings with swollen hollowed stem	Pulse stem fly	*Ophiomyia phaseoli*
2. Seedling stem cut and plant lying on ground	Tobacco cutworms, Black cut worms (Potato, cabbage)	*Spodoptera spp., Agrotis ipsilon*
3. Woody stem and young plants base covered with earth tube and wilting of plants	Termites (Sugarcane, many orchard crops and trees)	*Odontotermes Spp*

Estimation of Pest Population Density

Pest population studies are helpful in pinpointing the factors that bring about numerical changes in the natural population and also in understanding the functioning of the life system of the pest species.

Pest population estimation studies are of two types:

(i) Extensive studies: These studies are spread over a large area and are needed to understand the distribution pattern of a population, to predict the damage it is likely to cause, to initiate control measures and to relate changes in the population to certain climatic or edaphic factors. A particular area is observed once or at the most a few times during the season and counts are made of a particular developmental stage of the pest.

(ii) Intensive studies: These studies involve repeated observations in a given area when it is desired to determine the contribution of various age intervals to the overall rate of change in the population or the dispersal of species. In this case the numbers of successive developmental stages are counted, and life-tables and budgets are prepared for determining the key factor(s).

Measurements taken to estimate pest population density fall into three categories, viz. absolute estimates, relative estimates and population indices

1. Absolute estimates:

The total number of insects per unit area (1 ha, 1 m row length, 1 m^2 quadrat, etc.) is the absolute estimation. The numbers per unit of the habitat (per plant, shoot or leaf) indicate the density of population. The estimates of absolute population and population density are used for preparing life tables, study of population dynamics of field populations and to calculate oviposition and mortality rates.

The following methods are commonly employed for estimating absolute population:

(i) Quadrat method: Small areas or quadrats are chosen at random from a large area which contains the population. From a quadrat, the insects may be counted or collected directly as in the case of fairly immobile but relatively large insects such as cutworms, caterpillars and grasshoppers. In case of tissue borers such as sugarcane borers, maize borer, etc., the estimation is done by first removing the infested plants from the quadrats and then counting them after splitting open the plant parts.

(ii) Capture, marking, release and recapture technique: This technique is generally used for estimating the population of flying insects. The losses or gains in a population over a period can be determined with the help of this method.

The total population is estimated by using the following formula:

$P = N \times M/R$ Where P = Population of insects, N = Total number of insects caught, M = Number of marked individuals released, R = Number of marked individuals recaught

2. Relative estimates: In relative population estimates, the samples usually represent an unknown constant proportion of the population. A given amount of labour and equipment is utilised to yield much more data than is possible for absolute estimates.

Such estimates are useful in making comparisons in space or time. These are useful for studying the activity patterns of a species or for determining the constitution of a polymorphic population.

(i) Catch per unit time or effort: Various types of collection nets are available for use in different habitats and the sweep net is the most widely used for sampling insects from vegetation. Only those individuals on the top of the vegetation and those that do not fall off or fly away on the approach of the collector can be caught with the sweep net.

(ii) Line-transect method: If one walks in a straight line at a constant speed through a habitat, the number of individuals can be counted. This technique is used for quantitative comparisons both between different species, and between different occupiers of habitats. The number of organisms per unit area or their density can be calculated by the formula:

$D = Z/2R \ (V + \bar{W})^{1/2}$ where, D = Density, Z = Number of encounters between the observer and the organism in a unit time, R = Radial distance within which the organism must come in contact with the observer to affect an encounter, V = Average speed of the observer, W = Average speed of the organism

(iii) Shaking and beating: Some insects can be collected on ground by shaking or beating the plants. A piece of cloth or polythene may be laid out under the plants and the plants are vigorously shaken. The insects fall on the cloth or polythene and need to be counted immediately before they disperse. The gram pod borer, *Helicoverpa armigera* (Hubner) larvae may be sampled by vigorously shaking the chickpea plants.

(iv) Remote sensing: Remote sensing technology has long been used for monitoring insect infestation in field crops. It is based on the principle that the absorbance and reflectance of plants in response to pest attack changes and these changes are recorded by a device from far away. Remote sensing platforms can be aircraft, satellites or ground based. The remote sensing techniques include full-colour photography, infrared (IR) wavelength and multiband spectrometers.

3. Population indices: Population indices do not count insects at all, but rather they are measures of insect products or effects. Under field conditions, it is not possible to estimate the absolute population in most of the cases. It, therefore, becomes necessary to establish a relationship between absolute estimates and population indices or the relative estimates so that the latter two types of estimates could be converted to absolute terms by using certain correction factors.

(i) Insect products: In some cases, a species that is difficult to sample creates products directly that are easily sampled by absolute methods. The insect product most often sampled is frass or excrement of lepidopterous defoliators. The rate at which frass is produced can be estimated from the amount falling into a box or funnel placed under the trees. The size and shape of the frass pellets is rather constant for a given species and instar; this allows one to identify the species and age composition of defoliators.

(ii) Plant damage: The amount of damage caused by insects to crop plants is a function of the pest density, the characteristic feeding or opposition behaviour of the species and the biological characteristics of the plants. Different methods have to be adopted for measuring damage by direct and indirect pests

Question - Answer

MCQs

1. Little leaf of brinjal is caused by-

   a) *Aceria jasmini,*   b) *Cestius phycitis,*
   c) *Plutella xylostella,*   d) *Thrips tabaci*

   Ans. b)

2. Scientific name of boll weevil is-

   a) *Phenacoccus hirsutus*   b) *Anthonomous grandis*
   c) *Tectana grandis*   d) *Sylepta derogate*

   Ans. b)

3. Pest population studies are of-

   a) 3 types   b) 2 types
   c) 4 types   d) 1 type

   Ans. b)

4. Pest population estimates are of-

   a) 3 types   b) 2 types
   c) 4 types   d) 5 types

   Ans. a)

5. Relative estimate includes-

   a) Use of traps   b) Insect products
   c) Plant damage   d) All of the above

   Ans. a)

6. Choose the paddy pest from the options below

   a. *Sesamia inferens*   b. *Stenchaetothrips biformis*
   c. *Pelopidas Mathias*   d. *Chilo partellus*

   Ans. b

7. Choose the cotton pest from the options below.

   a. *Rhopalosiphum maidis,*   b. *Tetraneura nigriabdominalis*
   b. *Atherigona approximate*   d. *Sesamia inferens*

   Ans. i

8. Average depth of net trap is-

    s. 5-10 cm   b. 10-15 cm
    c. 20 cm   d. 25 cm

    Ans. c

9. How many types of sampling are there?

    a. 1   b. 2
    c. 3   d. 4

    Ans: b

10. DBM is a specific insect of............

    a. Brinjal.   b. Tomato.
    c. Cabbage   d. Bottle gourd.

    Ans. c

11. IPM is also known as..........................

    a. IPC.   b. IDM.
    c. IPA.   d. IIM.

    Ans. a

12. Polyphagous insect-pest is...............................

    a. Locust.   b. DBM.
    c. Mango hopper.   d. Green hopper.

    Ans: a

13. Monophagous insect-pest is ....

    a. Rice yellow stem borer.   b. Cotton jassid.
    c. Rice weevil.   d. All of the above.

    Ans: a

14. Which of the following describe *Trichogrmma*?

    (a). It is a parasite of larvae.   (b). It is an egg parasitoid.
    (c). Both a and b.   (d). It is a predator.

    Ans: b)

15. Vector of rice tungro virus is. ....

    a. GLH.   b. BPH.

    d. WBPH.   d. None of the above

    Ans: a

16. Vector of grassy stunt disease is ....

    a. GLH.   b. BPH.

    c. WBPH.   d. All of the above.

    Ans: b

17. Which of the following is Brown plant hopper?

    a. *Nephotettix nigropictus*   b. *Cofana spectra*

    c. *Empoascanara maculifrons*   d. *Nilaparvata lugens.*

    Ans. d

18. Which is the following pest attack sorghum crop on the lower surface of leaf blades mostly during morning hours

    a. Stem borer   b. Pink stem borer

    c. Sorghum shoot fly   d. Earhead bug.

    Ans. c

19. Scientific name of Sugarcane *Pyrilla* –

    a. Pyrilla borer   b. *Pyrilla perpusilla*

    c. *Chilo infuscatellus*   d. *Aleurolobus barodensis*

    Ans. b

20. Which following absolute estimate method is commonly employed for estimating absolute population?

    a. Quadrat method   b. Line-transect method

    c. Use of traps   d. Remote sensing.

    Ans. a

21. The total population is estimated by using the following formula –

    a. $P = N \times M/R$ b. $P = M \times N/R$

    c. $Q = N \times M/R$ d. $Q = M \times N/R$

    Ans. a

22. Which is a sucking pest?

    (a) *Thrips tabaci*   (b) *Holotrichia consanguinea*   (c) *Pericallia ricini*

    Ans: a

23. Which is a groundnut pest?

    (a) *A. moorei*   (b) *Scirpophaga excerptalis*   (c) *Sesamia inferens*

    Ans: a

24. Which is not a pulse pest?

    (a) *Helicoverpa armigera*   (b) *Perigaea capensis*   (c) *Lampides boeticus* .

    Ans: b

25. Dead heart in paddy is caused by

    (a) Green leafhopper   (b) Thrips   (c) Stemborer

    Ans: c

SAQs

1. What is pest identification?

Pest identification is the first and most important step in any pest management situation. Integrated pest management depends on "field scouting," or monitoring pest populations and crop development.

2. Why identification is important in pest management?

Correct identification of a pest species is the first step in scientific pest control. It provides a key to publish information on the life history, behaviour, ecology of the insects, and to other data important in the development of control measures.

3. How do pest cause damage?

Insects with chewing mouthparts, for example, grasshoppers, caterpillars, and beetles, cause feeding damage such as holes or notches in foliage and other plant parts, leaf skeletonizing (removal of tissue between the leaf veins), leaf defoliation, cutting plants off at the soil surface, or consumption of roots.

4. What is population estimate?

Population estimates can describe the total population size as well as demographic characteristics such as age, sex, or education level. Population estimates are dependent on the demographic components of change: mortality, fertility, and migration.

5. How do you estimate pest population?

The total number of insects per unit area (1 ha, 1 m row length, 1 m² quadrat, etc.) is the absolute estimation. The numbers per unit of the habitat (per plant, shoot or leaf) indicate the density of population

6. What is remote sensing?

Remote sensing technology has long been used for monitoring insect infestation in field crops. It is based on the principle that the absorbance and reflectance of plants in response to pest attack changes and these changes are recorded by a device from far away. Remote sensing platforms can be aircraft, satellites or ground based. The remote sensing techniques include full-colour photography, infrared (IR) wavelength and multiband spectrometers.

7. Example of two direct pests?

Bollworms on cotton and fruit borers in fruit and vegetable crops.

8. The number of organisms per unit area or their density can be calculated by which formula?

$D = Z/2R \ (V + \overline{W})^{1/2}$ Where, D = Density Z = Number of encounters between the observer and the organism in a unit time, R = Radial distance within which the organism must come in contact with the observer to affect an encounter, V= Average speed of the observer, W= Average speed of the organism

9. Examples of indirect pests?

Lepidopterous caterpillars, leaf beetles, grasshoppers, etc.

10. What is an absolute estimate?

The total number of insects per unit area (1 ha, 1 m row length, 1 m² quadrat, etc.) is the absolute estimation. The numbers per unit of the habitat (per plant, shoot or leaf) indicate the density of population. The estimates of absolute population and population density are used for preparing life tables, study of population dynamics of field populations and to calculate oviposition and mortality rates.

11. How can you define pest?

Derived from french word 'peste' and Latin word 'Pestis' meaning plague or contagious disease. Pest is any animal which is noxious, destructive or troublesome to man or his interest.

12. How can we categorise pests?

We categorise pest mainly based on three points that are based on i) infestation ii. Based on occurrence iii. Pest categories according to EIL, GEP and DB

13. How can we diagnose pest damage?

Insect and non-insect pests cause a particular type of damage to the plant parts often characteristic to particular pests. The pest mostly insect not present on the site of damage makes it difficult to know the causal agent. Sometimes, the symptoms of damage caused by insects may closely resemble to those resulted due to pathogens or due to nutritional disorders. So, practical experience makes familiar about the correct diagnosis on the basis of visual symptoms of damage to take appropriate control measures.

14. Which things we should remember before diagnose pest damage?

We should remember that-

» Don't be in too much of a hurry, slow down, cut open the plant and have a look inside.

» Use a hand lens to look for small insects and their stages.

» Talking to the farmers about the topic that we are researching.

15. What is light trap? What are the advantages and limitations?

It is the most widely used visual trap employed for sampling agricultural pests, particularly moths, hoppers and beetles, etc. A light trap essentially consists of a light source above a funnel and a container below to collect the catch. It is covered with a protective roof. The light source is generally an oil lamp or electric bulb or a fluorescent tube.

Advantages:

i.   Insect-pest monitoring to document seasonal dynamics of their populations in the agro-ecosystems for timely pest management.

ii.   Mass trapping of selective phototrophic insect-pests likes macro-lepidopteran insects viz., hairy caterpillar, semilooper etc.

Limitations:

If used alone, light traps may fail to collect important or infected vectors, and light traps are inefficient or ineffective when competing ambient light is present.

16. List out the different traps used for the sampling of pest population.

   i. Light trap
   ii. Adhesive trap
   iii. Flying insect trap
   iv. Terrestrial arthropod trap
   v. Aquatic arthropod trap
   vi. Bait trap
   vii. Water trap
   viii. Pheromone trap

17. What is indirect technique?

In indirect techniques, insect population are estimated by measuring the effect of insects on crop plants or through some indirect sampling methods called population indices. e.g. number of leaves mined by leaf miners, percentage defoliation by hairy caterpillars and armyworms, percentage of plant attacked, percentage of fruiting forms or bolls damaged, leafhopper injury grade in cotton, number of wilted or dead plants by termites, plants with "dead hearts" caused by borers, etc. The sampling technique involved with most of these measures is direct observation of results of insect injury. Sometimes, pest populations are assessed with the help of insect products which include larval and pupal skins (exuviae), frass, honey dew (e.g. secreted by aphid, whitefly) and nests of colonial insects. The measurement of frass drop in a collecting tray has been used to assess population size of several forest pests.

18. How can one identify Army worms?

Pale brown adults live for 1- 9 days and lay eggs singly in rows or in clusters on dry or fresh plants or on the soil. Freshly laid eggs are round, light green, turn pale yellow and finally black.

19. What are the main categories of pest estimation?

Measurements taken to estimate pest population density fall into three categories, viz. absolute estimates, relative estimates and population indices.

20. Mention any two factors which influence the relative estimates.

The relative estimations are influenced by following two factors- (a) Variation in behaviour of an insect with change in age. (b) Variation in level of activity of the pest as influenced by its diurnal cycle.

Fill in the blanks

1. Withering and drying of central shoot is known as_____

Ans. Dead heart

2. GPS stands for_____.

Ans. Global Positioning System.

3. Scientific name of Mango hopper is _____.

Ans. *Idioscopus spp.*

4. Red Borer is a pest of _____.

Ans. Coffee.

5. Scientific name of greenhouse pepper is _____.

Ans. *Capsicum annuum.*

6. Scientific instruments available for quantifying the extent of defoliation are_____.

Ans. Planimeters, photoplanimeters, leaf area meters etc.

7. Pest can be divided into .... pests based on level of incidence.

Ans. Endemic and epidemic

8. Rhizome weevil is a pest of ....

Ans. Banana

9. Full forms of GIS is ....

Ans. Geographical information system

10. The number of organisms per unit area or their density can be calculated by the formula_____.

Ans: $D = Z/2R (V + W)^{1/2}$.

11. The scientific name of Shoot fly is _____

Ans: *Atherigona approximate.*

12. *Holotrichia consanguinea*'s larva is_____ shaped.

Ans: fleshy 'C' - shaped.

True or false

1. Red boll worm is a pest of castor.

Ans. False

2. A dried up and scorched appearance of a plant is called " hopper burn "

Ans. True

3. Diamondback moth is a pest of mustard.

Ans. True

4. Webbed flowers and infested pods are symptoms of spotted pod borer attack.

Ans. True

5. Adults of green leafhopper have green with black spot and black patch on wings.

Ans. True

6. The relative estimates are influenced by variation in behaviour of an insect with change in age.

Ans. True

7. Population indices do not measure of insect products or effects, but they count insects.

Ans. False

8. The total population is estimated by using the following formula: $P = N \times M/R$.

Ans. True

9. Gram pod borer's pupa is white in colour.

Ans. False

10. In the total population formula $P = N \times M/R$, N refers to number of marked individuals released.

Ans. False

Chapter - 7

Identification of Bio-control Agents and Their Qualitative and Quantitative Estimation

Biological control is a process in which one species population lowers the number of another species by mechanisms such as predation, parasitism, pathogenesis or competition. Biological control involves use and manipulation of natural enemies by man.

Agents of Biological Control

1. **Predator:** A predator is a free living organism throughout its life which kills prey. It is usually larger than its prey and requires more than one prey to complete its development. e.g. green mirid bug, *Cytorhinus lividipennis* feeds mainly on the eggs and early stage nymphs of brown plant hopper in rice.

2. **Parasite / True parasite:** A parasite is an organism which attaches itself to the body of the other living organism either externally or internally and gets nourishment and shelter at least for a shorter period if not the entire life cycle. The organism which is attacked by the parasites is called host. Parasite is usually much smaller than its host and a single individual does not kill a host. e.g. *Apanteles taragame* is a parasite of the larvae of *Opisina arenosella*.

3. **Parasitoid:** A parasitoid is a special kind of parasite which is often about the same size as its host. It kills its host and requires only one host (prey) for development into free living adults. E.g. Braconids wasp is an egg larval parasitoid on cotton spotted bollworms, *Earias* spp.

4. **Pathogens:** Disease causing microorganisms or pathogens such as bacteria, virus, rickettsia, fungi, protozoa and nematodes can be used for controlling crop pests. When microorganisms or their products (toxins) are employed by man for the control of pests, it is referred to as microbial control. This is an extension of biological control where instead of macrobes (parasites and predators), microbes (pathogens) are employed. E.g. *Bacillus thuringiensis* is widely used against Lepidoptera, Coleoptera and Dipteran insects.

Coleoptera (Coccinellids, Carabids, Staphylinids)

1. Coccinellidae (Lady bird beetles)

- » Adults are small, oval or spherical, convex, brightly covered with coloured spots.
- » Head concealed in pronotum.
- » Tarsi 4 segmented and 3rd segment concealed in the deeply bilobed second segment.
- » Antennae short, clubbed 3 to 6 segmented.
- » Larvae campodeiform usually covered with minute tubercles or spines.
- » Both adults and grubs are predaceous on soft bodied insects and mites. Example: *Coccinella septempunctata*

2. Carabidae (Ground beetles)

- » They live under stones, ground and bark.
- » Oval and broad, dorsoventrally flattened beetles with metallic, bright dark coloration.
- » The elytra are firmly attached together and wings may be atrophied.
- » The legs are slender and adapted well for running and digging.
- » Adults and larvae are carnivorous; most of them feed on caterpillar. Example: *Anthia sexguttata, Ophionea indica.*

3. Staphylinidae (Rove beetles)

- » Small to medium sized insects.
- » Short elytron, considerable portion of abdomen is exposed.
- » The mandibles are well developed and cross in front of the head.

» The well developed hind wings are kept folded under the elytra at rest.

» Predaceous on saprophagous insects. Example: *Paederus fuscipes*

Neuroptera (Chrysopids)

Chrysopidae (Ant lions)

» Bright green body with wing veins and iridescent eyes.

» Antennae filiform.

» R_3 arising from main separating from R_1 and does not fuse distally with Sc.

» They (Larva) commonly feed on aphids, thrips, psyllids, coccids, jassids etc. Example: *Chrysoperla carnea*

Hemiptera (Mirids, Reduviids, Pentatomids)

Miridae

» These are delicate small to medium sized elongated insects.

» Antennae and rostrum are 4 segmented.

» Ocelli absent.

» Tarsi 3 segmented.

» Distinct cuneus and an indistinct embolium.

» Ovipositor well developed in females. Example: *Psallus spp.*, *Cyrtorhinus lividipennis*

Reduviidae (Assassin bugs)

» Brown to black medium sized insects.

» Rostrum 3 segmented, curved and its tip fits in a posterior groove.

» Fore legs bear adhesive pads on tibia.

» Metathoracic scent glands absent.

» Most of these are predaceous and some are blood sucking. Example: *Rhynocoris marginatus*

Pentatomidae (Stink bug)

» Brightly coloured, medium sized, shield shaped and produce disagreeable odour.

» Antennae 4 or 5 segmented and base is concealed by lateral margin of the head.

» Scutellum triangular and extends posterior and covers the wings entirely.

» Hemelytra well developed with 5-12 veins.

» Mostly phytophagous and some are predaceous.
Example: *Cantheconidia sp.*

Diptera (Syrphids, Tachinids)

Syrphidae (Hover flies)

» Adults are often seen hovering or nectaring at flowers, hence called hover flies.

» Wasp or bees like, brightly coloured with yellow stripes or bands.

» Larvae are insectivorous and prey on aphids, thrips and other plant sucking insects.

» Hover flies are distinguished from other flies by a spurious vein, located parallel to the forth longitudinal wing vein. Example: *Syrphus confractor*

Tachinidae (Tachinid flies)

» These are stout flies, medium to large, that are very bristly, particularly around the posterior of the abdomen.

» Adults with pteropleural and hypopleural bristles.

» Abdomen clothed with long conspicuous marginal, apical and dorsal bristles.

» The larvae are internal parasitoids of insects especially Lepidoptera and adults feed on nectar.

» Pupates outside the body of the host.
Example: *Sturmiopsis inferens*

Lepidoptera (Pyralid – *Epiricania melanoleuca*)

Epipyropidae

» Small moths.

» Larvae are ectoparasites, the hosts typically being fulgoroid planthoppers.

» Also called as planthopper parasite moths.
Example: *Epiricania melanoleuca.*

Hymenoptera (Trichogrammatids, Eulophids, Scelionids, Mymarids, Braconids, Ichneumonids)

Trichogrammatidae

» These are very minute insects with three segmented tarsi.

» Fore wings are broad with rows of microscopic hairs.

» These parasitize the eggs of large number of insects especially Lepidoptera.

» They are not strong fliers and passive fliers (move along the wind).

» Forewings are stubby and paddle shaped, with along fringe of hinged setae around the outer margin to increase the surface area during the down stroke.
Example: *Trichogramma chilonis*

Eulophidae

» They are minute pupal parasites.

» Forewing is narrower with pubescence on the wing lamina.

» Hairs are not arranged in rows.

» Ovipositor is present almost at the tip of the abdomen.
Example: *Tetrastichus israelli*

Scelionidae

» These are small insects.

» These parasitize the eggs of Lepidoptera, Orthoptera, Dictyoptera, Hemiptera etc.

» The phenomenon of phoresy is observed.
Example: *Telenomus beneficiens.*

Myramidae

» Body usually 1.5 mm or less in length.

» Scutellum normally divided transversely into an anterior and posterior part.

» Fore wing with venation greatly reduced. Hind wing stalked or petiolate, the membrane of the disc usually originating apically from the stalk or rarely with stalk only.

» Tarsi 4-segmented (or) 5-segmented.
  Example: *Gonatocerus similis*

Braconidae

» Most braconids are primary parasitoids (both external and internal) on other insects.

» Very small stout bodied insects.

» Ovipositor is long and many species are egg-larval parasitoids.

» Only one recurrent vein, cross vein 2m-Cu is absent in the fore wing.

» Pupation within cocoon inside or outside the body of the host.
  Example: *Bracon hebetor.*

Ichneumonidae

» These are important parasitoids of larvae and pupae of Coleoptera, Hymenoptera and Lepidoptera.

» Antennae are long with typical 16 or more segments.

» Ovipositor is long and extruded permanently.

» Long antennae, presence of trochantellus on hind femur.

» Absence of anal lobe of hind wing.

» Distinct pterostigmata and narrow costal cell on forewing.

» Two distinct recurrent cross veins.
  Example: *Isotima javensis.*

Spiders (Arachina)

Spiders are not insects, although they are closely related, but they are true friends as predators against number of insect pests. Spiders are arachnids and have two main body parts (instead of three) and eight legs (instead of six).

Some spiders spin webs in which they patiently wait for their prey to become entangled, others actively hunt down their prey, and still others sit motionless on plants or flowers and pounce in ambush on unwary insects that wander near. Hunting spiders have prominent eyes and good eyesight to see their prey, and instead of large webs they construct small silken shelters in which to rest. Web spinners create silken webs, but have poor vision and rely on the vibrations of insects captured in their webs to detect their prey. Ambush, or sit-and-wait, spiders have patience on their side. No matter what method they use, though, spiders substantially reduce populations of pest insects in the garden and crop field.

Example: Web spinners such as orb weavers; hunting spiders, including wolf and jumping spiders, ambush spiders such as funnel weaver (or grass), crab (or flower) spiders, and daddy longlegs (a distant relative of spiders).

Mites (Phytoseiids)

Phytoseiidae

The Phytoseiidae are a family of mites which feed on thrips and other mite species. They are often used as a biological control agent for managing mite pests. Because of their usefulness as biological control agents, interest in phytoseiids has steadily increased over the past century.

Description and life history

There are several families of predatory mites, but the phytoseiids are the most important for agriculture and horticulture because they are voracious predators of spider mites. Many of these predatory mites are commercially available. They live in the soil or leaf litter. They feed with a pair of needle like chelicerae.

Prey species: Although these mites almost always prey on other mites and small insects, many can also feed on honey dew or pollen during times of prey scarcity. Many phytoseiid mites are generalists, but a few have specific prey requirements. These are very effective control agents due to their short generations, high fecundity and hearty appetites.

Example: *Phytoseilus persimilis* and *Amblyseius californicus*

Weed feeding herbivores – *Zygrogramma bicolorata*

Zygogramma bicolorata, variously referred to as the Parthenium beetle or Mexican beetle, is a species of leaf beetle in the family Chrysomelidae under order Coleoptera. *Z. bicolorata* is a small lead beetle with a brown head, brown and yellow graduated pronotum and yellow elytra marked with characteristic elongated brown stripes. The pattern on the elytra is greatly variable. *Z. bicolorata* is native to Mexico, but has been introduced to parts of India and Australia. Adults and larvae are used as a form of biological pest control in India in order to control invasive *Parthenium hysterophorus*.

Eggs are generally laid on the ventral surface of both young and old leaves, and occasionally on the upper surface of leaves, stems and flowers of host plants. Eggs are yellow to orange, elongate cylindrical or oblong with fine reticulations on the surface. The eggs hatch in 4–5 days. Larvae are pale yellow, turning white as they grow, feeding for 10 to 15 days on leaves whilst growing through four instar stages. On maturity the larvae enter the soil and pupate below up to 15 cm depth. The total life cycle of the beetle is just over 100 days.

Qualities of successful parasitoid in biological control programmes

»   Should be adoptable to environmental conditions in the new locality.

»   Should be able to survive in all habitats of the host.

»   Should be species-specific.

»   Should be able to multiply faster than the host

»   Should have more fecundity

»   Life cycle should be shorter than host

»   Should have high sex ratio

»   Should have good searching capability for host

»   Should be amenable for multiplication in labs

»   Should have quick dispersal capability

»   Should be free from hyperparasites

»   Should not be the parasite on productive insects

Qualities of predator:

»   Narrow host range: Generalized predators may be good natural enemies but they don't kill enough pests when other types of prey are also available.

»   Climatic adaptability: Natural enemies must be able to survive the extremes of temperature and humidity that they will encounter in the new habitat.

» Synchrony with host (prey) life cycle: The predator should be present when the pest first emerges or appears.

» High reproductive potential: Good bio-control agents produce large numbers of offspring.

» Efficient search ability: In order to survive, effective natural enemies must be able to locate their host or prey even when it is scarce. In general, better search ability results in lower pest population densities.

» Short handling time: Natural enemies that consume prey rapidly or lay eggs quickly have more time to locate and attack other members of the pest population.

» Survival at low host (prey) density: If a natural enemy is too efficient, it may eliminate its own food supply and then starve to death. The most effective biocontrol agents reduce a pest population below its economic threshold and then maintain it at this lower equilibrium level.

» Efficient eating ability: They should kill or consume much prey.

» All should be predatory: Males, females, immatures, and adults may be predatory.

» Should prey on all host stages: They attack immature and adult prey.

Characteristics of ideal microbial bio-control agent:

1. Non-pathogenic

2. Broad spectrum of activity

3. Fast growth and Sporulation ability

4. Cultured under artificial media

5. Capable of abundant production

6. Economically viable

7. Shelf life

8. It must be efficacious under different environment conditions

9. Compatible with fertilizer and non-toxic to beneficial microbes

10. Formulations are easier and methods of application are beneficial

11. Good persistence and survival capacity

12. Biologically competitive with other microbes.

Biological parameters assessed in control quality criteria of natural enemies

i.   **Quantity:** Number of live natural enemies in container

ii. **Sex ratio:** Minimum % females

iii. **Emergence:** Emergence rate to be specified for all organisims sold as eggs or pupae.

iv. **Fecundity:** Number of offspring produced during a certain period

v.   **Longevity:** Minimum longevity in days.

vi. **Parsitism:** Number of host parasitized during a certain period.

vii. **Predation:** Number of prey eaten during a certain period.

viii.**Adult size:** hind tibia length of adults, sometimes pupal size (size is often a good indication for longevity, fecundity and parasitization/predation capacity).

Criteria to be added in near future:

i.   **Flight:** short- or long-range flight capacity

ii. **Field performance:** capacity to locate and consume prey or parasitize hosts in crop under field conditions

Comments:

»   Quality control is done under standardised test conditions of temperature (usually 22 ± 2o C or 25 ± 2o C), relative humidity (usually 75 ± 10 %) and light regime (usually 16 L: 8 D), that are specified for each test.

»   All numbers / ratios / sizes should be mentioned on the container or packaging material

»   Fecundity, longevity and predation capacity tests can often be combined

»   Expiration date for each shipment should be given on packaging material

»   Guidelines should be usable for all product formulations

Maintenance of *Trichogramma* quality

The following practices are required to maintain the quality of *Trichogramma*.

»   Production colonies should be periodically replaced with individuals from a stock culture maintained on the natural or target host.

» Producers should also periodically assess the percent host egg parasitisation, adult emergence and the sex ratio of emerged adults to be sure that they are within acceptable standards.

» Standards for established cultures on *Corcyra* are 95±5 % egg parasitisation. 90±5 % adult emergence and a sex ratio of 1 to 1.5 female per male.

Impact assessment of natural enemies

There is a pressing need for measuring impact of natural enemies on insect pest management. There are different quantitative and qualitative methods for impact assessment of natural enemies. Qualitative techniques include direct observation, correlation, and seasonal abundance; Quantitative techniques include laboratory evaluation, field cage evaluation and field evaluation.

Qualitative methods

Direct observation in the field

Direct observation provides important information related to natural enemy -prey relationships.

Correlation

It is carried out by monitoring populations of pest and natural enemy during different seasons and correlating numbers of increase of the prey and number of natural enemy.

Seasonal abundance of natural enemy

Seasonal and relative abundance of natural enemy are to be monitored to assess their seasonal variation.

Question - Answer

MCQs

1. What are the dragonflies use to get rid of –

    a) Aphids   b) Mosquitoes   c) Earthworms d) Honeybee.

    Ans: Mosquitoes

2. Who used the term "Biological Control" first?

    a) Anderson in 1885   b) Locatelli in 1909   c) Smith in 1919

    Ans: c

3. What are biocontrol agents for controlling butterfly caterpillars?

    a) *Bacillus thuringiensis*   b) *Lactobacillus*
    c) *Acetobacter aceti*   d) Treponema palladium

    Ans. a

4. From the following which is not a Predator?

    a) Lady Bird Beetle   b) Dragon Fly
    c) Blister beetle   d) Honey bee

    Ans. d

5. Praying mantis is used in biological control of insect pests as a

    (a) Pathogen   (b) Parasite
    (c) Predator   (d) Parasitoid

    Ans. c

6. NPV stands for –

    (a) Nuclear Polyhedrosis Virus,
    (b) Nuclear Polyhedrosis Viroid,
    (c) Nuclear Pathogenic Viroid,
    (d) Nuclear Pathogenic Virus

    Ans. a

7. Which one of the following is a parasitoid

    (a) Robber fly,   (b) Giant water bug,
    (c) Ladybird beetle,   (d) *Trichogramma sp.*

    Ans. d

8. Which one is the characteristic feature of Ladybird beetle –

    (a) Only adults are predator   (b) Antenna is 6-11 segmented

    (c) Tarsi is 4 segmented   (d) Both b and c.

    Ans. d

9. *Vespa cincta* is an insect of order –

    (a) Hymenoptera   (b) Lepidoptera

    (c) Hemiptera   (d) Dictyoptera

    Ans. a

10. Which is/are not the characteristics of ideal biocontrol agent?

    a) economically not viable

    b) narrow spectrum of activity;

    c) fast growth and Sporulation ability

    d) didn't culture under artificial culture.

    Ans. c

12. Which is not a biochemical method of identification?

    a) Staining   b) Starch hydrolysis

    c) Catalase test   d) Nitrate reduction test

    Ans. a

13. Order of lady Bird beetle is

    a) Lepidoptera   b) Coleoptera

    c) Hemioptera   d) Dictyoptera

    Ans. b

14. Family of *zygrogramma* –

    a) Trichogrammatidae   b) Coccinellidae

    c) Chrysopidae   d) Chrysomelidae

    Ans. d

Fill in the Blanks

1. An example of entomopathogenic bacteria is ....

     Ans. *Bacillus thuringiensis.*

2. Bracon hebetor is a ....parasitoid of Indian meal moth.

     Ans. Larval

3. *Trichogramma australicum* is used against *Chilo suppressalis* as an ....

     Ans. egg parasitoid.

4. *Anax guttatus* is an insect under Family .... and Order ....

     Ans. Aeshnidae and Odonata.

5. Aphids can be controlled by ....

     Ans. Hover flies (*Dideopsis segrota*).

6. Familiy of aphid lion is ....

     Ans. Chrysopidae.

7. .... is used to Control of *Lentana camara*.

     Ans. *Crosydosema lentani*

8. *Trichogramma australicum* is used against ....

     Ans. *Chilo suppressalis.*

9. Order of *Epiricania melanoleuca* is ....

Ans. Lepidoptera.

True or False

1. Egg parasitoid is an example of inundative release (True)

2. Parasitiods is Diurnal. (True)

3. Fungi enter the host cell via Integument (True)

4. There are 5 kinds of predatism (True)

5. When same species parasite attacks on host cell then it is termed as Hyper Parasitism. (False)

6. Bio-control organization of India is situated in Kerala - (False)

7. Fungal strains were stored on PDA medium at 4 degree centigrade - (True)

8. In the qualitative estimation of *Trichoderma* pH was adjusted to 7.2 - (True)

9. NPV belongs to the sub group Baculovirus. (True)

10. Mantids have prognathous head. (True)

11. *Microvelia atrolineata* feeds on Brown Plant Hopper and Onion Thrips. (False)

12. Antlion is a Neuropteran insect. (True)

13. The scientific name of cotton bollworm is *Helicoverpa punctigera*. (False)

14. *Beauveria bassiana* is an entamopathogenic fungi.(True)

15. Isolation and purification is method of identification of fungi. (True)

16. Neuroptera is the order of lady bird beetle. (False)

17. Infection with TMV virus is known as wilting disease. (False)

18. Appetite loss is a symptom of Appressoria. (True)

SAQs

1. What are the steps of "Classical Biological Control"?
The steps are 1) Exploration 2) Quarantine 3) Importation 4) Mass Production 5) Release and follow up

2. Define Virion
Complete Virus particle that consists of an RNA or DNA core with a protein coat.

3. What are the Macrobes and Microbes in Biological Control?
Parasitic and Predatory insects are under Macrobes and Pathogens are under microbes.

4. By which part Fungi enters the host cell?
Integument.

5. What is Appressoria?
In favourable conditions, the conidium germinates into a short germ tube which

gives out small swellings called appressoria

6.   Why is a fungus used as a biocontrol agent?

Fungal biocontrol agents do not harm the environment and have proven themselves to be a cheap alternative to harmful chemical pesticides, and they also do not need to be ingested by the host but can invade them directly. They are an effective biological method of pest control.

7.   What are biocontrol agents?

Biological control is a method of controlling pests which includes mites, insects, weeds and organisms causing diseases in plants. The biocontrol method is often an essential part of an integrated pest management program. The classical process of Biocontrol involves the use of a natural enemy or predators or parasitoids, where a large population of predators is released to achieve control over the pest population. In the inoculative method, measures are adopted to control the level of natural enemies through regular administration. Pathogens, parasitoids, competitors are some of the methods by which the control of pests is brought about. Predators, Parasitoids, entomopathogens and herbivores are used as biological control agents in many cases.

8.   What are some examples of biocontrol agents?

Biocontrol agents are of various types which have their advantages and disadvantages. Predators are those who consume insects such as free-living species. Lady Beetles are examples of this category. The larvae of hoverfly eat aphids and are readily used. Parasitoids are organisms which lay eggs in the body of insects and thus help in controlling the pest population. Wasps are examples. The pathogenic microorganisms like bacteria and virus are also used in the process of Biocontrol. Bacteria attack the digestive tracts of insects, and an example of a bacteria used in Biocontrol is *Bacillus thuringiensis* which is a soil-dwelling bacterium.

9.   Why are biocontrol agents used in pest control?

Use of Biocontrol agents are also called biological control and are mainly used for the reduction of pest population and produce yields which are free of any pests. The biocontrol methods are a long-term method and most importantly, a self-sustaining one which helps in the control of invasive cspecies of plants.

10. Which character is more important when choosing a biological control agent as perfect candidate?

i.   Narrow host range - Generalized predators may be good natural enemies but they don't kill enough pests when other types of prey are also available.

ii. Climatic adaptability - Natural enemies must be able to survive the extremes of temperature and humidity that they will encounter in the new habitat.

11. What is Aphiline?

Aphiline refers to a braconid wasp which is used in controlling the aphid species from damaging crops. Aphiline works with the use of stinging and parasitizing the comparatively small aphid species such as the peach aphid and the cotton aphid.

12. Name two advantages of microbes as biocontrol agents.

  1. These are cost-effective.
  2. These biocontrol agents reduce the use of chemicals and other pesticides.

13. Name two disadvantages of microbes as biocontrol agents.

  1. The high specificity against the target disease and pathogen may require multiple microbial pesticides.
  2. It may affect product quality.

14. What are Parasitoids?

The Parasitoids are a kind of biocontrol agent which lay eggs in the body of their hosts which ultimately leads to the death of the host. The dead host is then used by the larvae as a food source as is one of the most prevalent methods of biological control.

15. What is the full form of PGPR?

PGPR stands for plant growth promoting regulator

16. What is bio-control?

Bio control can be simply defined as the application of one living organism to control another. It is a method of controlling pests such as mites, weeds and plant diseases by using other organisms.

17. Write down the types of bio-control agents?

a. parasitoids b. predators c.entomopathogens- i.bacteria ii. Baculovirus iii.entomo-fungi iv.protozoans v.entomopathogenic nematodes.

18. Give an example of protozoan pathogen.

Nosema locustae against grass hoppers

19. Write the names of bacteria used for biological control.

Bacillus spp. and *Pseudomonus spp.*

20. What is parasitoid?

A parasitoid is a special kind of parasite which is often about the same size as its host. It kills its host and requires only one host (prey) for development into free living adults.

21. What is the scientific name of Indian Flower Mantis and it is used against which insect?

The scientific name of Indian Flower Mantis is *Creobroter pictipennis*. It is used against various species of grasshoppers, caterpillars and butterflies.

22. Cite an example of fungi and its target organism in biological control.

Green muscardine fungus (*Metarhizium anisopliae*) is used to control Coconut Rhinoceros Beetle (*Orycytes rhinoceros*).

23. In which purpose, the bacterial isolates are screened ?

The isolates are screened for phosphate solubilization activity, indole acetic acid (IAA) production and antifungal activity.

24. Full form of CLCuV.

Full form of CLCuV is Cotton leaf curl virus.

25. What is commonly used to reduce Cotton leaf Curl virus?

Microorganisms isolated from the rhizosphere, endosphere or phyllosphere are being commonly used to reduce disease

26. Write a brief note about Selection of bacterial strains as biocontrol agents.

The obtained bacterial colonies were screened for their potency as BCAs. Phosphate solubilization, indole acetic acid (IAA) production and antifungal activity were the parameters selected for screening.

27. Define IAA production.

To determine whether IAA is produced, 50 mL of LB broth was inoculated with 500 µL of 24 h old bacterial cultures. Incubation was done at 32 ± 1 °C at 180 r min–1 for 24–48 h in a shaking incubator. Following incubation, 1 mL of the culture was mixed with 2 mL of Salkowsky reagent and colour development was observed as described by Loper and Schroth.

28. Which fungal are needed for bio control assay ?

Fungal cultures of *Fusarium solani, Fusarium oxysporum* and *Rhizoctonia solani* is needed for bio control assey

29. What are the techniques of biological control?

Biological control involves three major techniques- A. Introduction B. Augmentation C. Conservation.

30. Mention two identifying characters of chalcidoidea?

    a. Possess long and filiform antennae.

    b. Wings are veined.

31. Mention the role of bio-control agents in IPM.

Biocontrol plays important role in IPM today with following aspects. (1) Biological control is less costly and cheaper than any other methods. (2) Biocontrol agents give protection to the crop throughout the crop period. (3) They do not cause toxicity to the plants. (4) Application of biocontrol agents is safer to the environment and to the person who applies them. (5) They multiply easily in the soil and leave no residual problem.

32. Write down the names of the super families of Hymenoptera.

Three super families of Hymenoptera are- A. Ichneumonoidea B. Chalcidoiea C. Bethyloidea.

33. Write down the merit and demerits of bio-control agents.

Merits of bio-control agents-

    » The biological control agents are environmentally friendly and cause no side effects.

    » Less cost compared to other Agrochemicals – pesticides and insecticides.

    » Easily available, easy to use and is effective throughout the season.

    » Helps in reducing the use of chemicals and other pesticides.

Demerits of bio-control agents-

    » It affects the product quality.

    » Pest is not completely destroyed by these biological control agents.

    » It is effective only for large scale

34. Write down the brief history of bio-control agents.

> » 900 A.D. – First time use of insect predators, red ants (*Oecophylla smaragdina*) by Chinese growers to control leaf chewing insects.

> » 1762- Introduction of Indian mynah bird, Gracula religiosa from India to control red locust, *Nomadacris septemfaciata.*

> » 1888- First successful control of insect cottony cushion scale (Icerya purchasi) a pest of citrus in California by using a predator, Vedalia beetle (*Rodalia cardinalis*) on large scale.

> » 1898- Predator, *Cryptolaemus montrouzieri* introduced from Australia into South India for control for citrus mealy bug.

> » 1929- *Rodalia cardinali*s introduced into India (TN) for control of insect cottony cushion scale (*Icerya purchasi*). Sources: Reddy, D.S. (2014) Applied entomology

> » 1937- *Aphelinus mali* introduced in to Coonor (TN, India) from North America to control apple wooly aphid, Eriosoma lanigerum.

> » 1960- *Spogosia bezziana* (Tachinid parasitoids) introduced from Srilanka into India for control of black headed caterpillar, Opisina arenosella

> » 1979- Eswarmoorthy and David first reportd granulosis visus infection of *Chilo infuscatellus.*

> » 1981- India first private insectary, Biocontrol Research laboratory was established at Banglore. Bio-control Organization in India. Nation Bureau of Agriculturally Important Insects [Formely Project Directorate of Biological Control (PDBC)] was established in 1993 at Bangalore, Karnataka, India.

Chapter - 8

Label and Toxicity
of Insecticides

Labels of Insecticides

Labels are legal documents providing directions on how to mix, apply, store, and dispose of a pesticide product. This means using a pesticide in a manner inconsistent with its labelling is a violation of federal law. The label is the manufacturer's main way to give the user information about the product. Both the label and leaflet are statutorily under the Insecticide Act, 1968.

The following information must be furnished on the label:

>   » Name of the pesticide (brand name, trade name, common name)
>   » Product Type
>   » Registration Number
>   » Establishment Number
>   » Manufacturer Name and Address
>   » Kind and name of active ingredient and their percentage
>   » Net Contents
>   » Hazards to Humans and Domestic Animals
>   » Directions for use
>   » Date of manufacture

» Expiry date

» Antidote statement

» Warning symbols and signals (warning symbol is of diamond-shaped consisting of two triangles (with a colour in the lower triangle and a signal in the upper triangle).

The leaflet must furnish the following:

» Name of the pests, weeds and disease against the chemical may be used.

» Direction of use

» Warning and cautioning statement, symptoms of poisoning, antidotes and first aid.

» Direction for storage, careful handling and method of disposal.

Symbols and signal words mean

The symbols and signal words on the pesticide label give you some quick information about the acute toxicity of the product

Four important symbols and words that show the potential hazards of pesticides - The hazard symbol will always appear inside one of the shapes shown below. These shapes and their warning words tell you the degree of hazard of the pesticide.

Toxicity of Insecticides

Toxicity: The toxicity of a pesticide is its capacity or ability to cause injury or illness. The toxicity of a particular pesticide is determined by subjecting test animals to varying dosages of the active ingredient (a.i.) and each of its formulated products. Insecticide toxicity can be either acute or chronic. Acute toxicity refers to injury or death after a short and often intense exposure of an organism to an insecticide. In laboratory toxicity testing, the duration of an acute exposure to an insecticide generally ranges from several to 48 hours.

Toxicity parameters

Toxic interaction of a chemical with a given biological system is dose related. The toxic effects are computed by probit analysis given by Finney in 1952. Toxicity of a given chemical to an organism can be measured using various parameters as listed below.

Median Lethal Dose (LD$_{50}$)

It is a unit to calculate the acute oral or dermal toxicity of an insecticide on a test animal (usually rat). It is referred as the minimum quantity of insecticide per unit weight of the organism required for killing 50% population of the test animal. It is generally expressed in milligrams per kilogram (mg/kg) of the body weight and for insects, microgrammes per gramme of the insect's weight (μg/g). The higher the LD$_{50}$ value, the lesser is the toxicity of the insecticide. The practical use of the LD$_{50}$ values is to know the comparative toxicities of various insecticides which are helpful both in the formulation of insecticides and in taking precautions while handling them. Dermal LD$_{50}$ is values higher than the oral ones and are more important to those who handle formulations or are likely to come in contact with insecticides.

Median Lethal Concentration (LC$_{50}$)

It is the concentration of insecticide in the external medium required to kill 50% of the test population. It pertains to fluid medium where the exact dose given to the insect can't be determined. This is used when the exact dose per insect is not known, but the concentration is known. This is expressed as the percent of active ingredient of the insecticide (percentage: 1/100) or as parts per million (ppm: 1/1,000,000).

Median lethal time (LT$_{50}$)

LT$_{50}$ is defined as the time required for killing 50% of the population at a certain dose or concentration. LT$_{50}$ expressed in hours or minutes.

Median knockdown dose (KD$_{50}$)

Dose of insecticide required to kill 50% of the insects.

Median knockdown time: (KT$_{50}$)

Time required to knock down 50% of the insects. KD$_{50}$ and KT$_{50}$ are used for evaluating synthetic pyrethroids against insects.

Median effective dose (ED$_{50}$) and Median effective concentration (EC$_{50}$)

ED$_{50}$ and EC$_{50}$ are defined as the dose or concentration of the chemical (IGR) required affecting 50% of population and producing desired symptoms in them. These terms are used to express the effectiveness of insect growth regulators (IGR).

Toxicity terms used to express the effect on mammals

» Acute toxicity: It refers to the ability to do systemic damage as a result of one-time exposure to relatively large amounts of the chemical. Toxic effect produced by a single dose of a toxicant, generally of short duration.

» Chronic toxicity: Toxic effects produced by the accumulation of small amounts of the toxicant over a long period of time.

» Oral toxicity: Toxic effect produced by consumption of pesticide orally.

» Dermal toxicity: Toxic effect produced when insecticide enters through skin.

» Inhalation toxicity: Toxic effect produced when poisonous fumes of insecticide are inhaled (fumigants).

» Other terms: Acute oral (through the mouth), acute dermal (through theskin), acute inhalation toxicity (through the lungs or respiratory system) etc.

5. Based on Toxicity

Insecticide toxicity is generally measured using LD_{50} (median lethal dose) – the exposure level that causes 50% of the population exposed to die.

Classification of the Insecticides	Medium lethal dose by the oral route (acute toxicity) LD_{50} mg/kg.	Medium lethal dose by the dermal route (dermal toxicity) LD_{50} mg/kg.	Colour of identification band on the label in lower triangle	Symbols and word in upper triangle
Extremely toxic	1-50	1-200	Bright red	POISON, skull with cross-bones
Highly toxic	51-500	201-2000	Bright yellow	POISON
Moderately toxic	501-5000	2001-20000	Bright blue	DANGER
Slightly toxic	More than 5000	More than 20000	Bright green	CAUTION

Categorisation of pesticides

Depiction	POISON (1)	POISON (2)	DANGER	CAUTION
Colour of lower triangle	Bright red	Bright yellow	Bright blue	Bright green
Toxicity class	Extremely toxic	Highly toxic	Moderately toxic	Slightly toxic
Oral LD_{50} value (mg/kg)	<50	51-500	501-5000	>5000
Signal words (Upper half)	POISON (in red)	POISON (in red)	DANGER	CAUTION
Warning words (Outside the diamond)	Keep out of reach of children, if swallowed or symptoms of poisoning occur, call doctor.	Keep out of the reach of children.	Keep out of the reach of children.	---

Studies on insecticides of different groups – their common names, trade names, colour symbols, signal and toxicity levels, Studies on different insecticides formulation and consideration for their specific uses, Consideration for selection of insecticides against crop pests

1. Common name 2. Formulation 3. Trade name 4. Colour symbols 5. Toxicity class 6. Signal word 7. Group (on the basis of chemical nature) 8. Group (on the basis of mode of entry) 9. Group (on the basis of mode of action) 10. Formulation Dose (ml or g/lit. of water) 11. Target pest 12. Target crop

1	2	3	4	5	6	7	8	9	10	11	12
Abamectin	1.90 % EC	Abba	Yellow	Highly toxic	Poison	Avermectin	Contact and Stomach	Nerve poison	0.75 ml	Mites	Rose
Acephate	75 % SP	Starthane, Asataf	Blue	Moderately toxic	Danger	Organo phosphate	Contact & Systemic	Nerve poison	0.75 – 1.5 gm	Jassid, Bollworm, Aphid, Yellow stem borer, Leaf folder, Plant Hoppers, Green leaf hopper	Cotton, Sunflower, Rice
Acephate	95 SG	Hunk	Blue	Moderately toxic	Danger	Organo phosphate	Contact & Systemic	Nerve poison	1 g	Stem borer, Leaf folder, Brown plant hopper, Jassids, Thrips, Fruit borer, Aphid	Rice, Cotton, Chilli
Acephate	97 % DF	Perito Ultimos	Blue	Moderately toxic	Danger	Organo phosphate	Contact & Systemic	Nerve poison	1-1.5 g	Jassids & Boll worm complex, Yellow stem borer, Leaf folder, Plant hoppers, Green leaf hopper	Cotton, Paddy
Acetamiprid	20 % SP	Pride, Tata Manik	Yellow	Highly toxic	Poison	Neonicotinoids	Systemic	Nerve poison	0.2 g	Aphid, Jassid, Whitefly, Thrips, Brown Plant Hopper	Cotton, Cabbage, Okra, Chilli, Rice
Afidopyropen	50 g/L DC	Sefina	Blue	Moderately toxic	Danger	9D group of pyropene chemistry	Systemic	Silencing of TRPV (transient receptor potential vanilloid)	2 g	Whitefly and Jassid	Brinjal, Cotton, Cucumber
Alpha cypermethrin	10 EC	Stop	Yellow	Highly toxic	Poison	Synthetic pyrethroid	Contact & Stomach	Nerve poison	1 ml	Boll worm	Cotton

Insecticide	Formulation	Trade Name	Colour	Toxicity	Poison	Class	Mode	Action	Dose	Pest	Crop
Alpha cypermethrin	10 SC	Shivalik	Yellow	Highly toxic	Poison	Synthetic pyrethroid	Contact & Stomach	Nerve poison	1 ml	Boll worm	Cotton
Aluminum Phosphide	56 %	Phostoxin	Yellow	Nerve poison	Poison	Inorganic	Respiratory	Nerve poison	3 g per ton	Rice Weevil (*Sitophilus oryzae*), Lesser Grain Borer, Khapra Beetle (*Trogoderma granarium*), Rust Red Flour Beetle, Saw Toothed Grain Beetle, Caddle Beetle, Drug Store Beetle, Cigarette Beetle, Pulse Beetle	Stored grains, fruits and process products
Azadirachtin (Neem Seed Kernel Based)	0.15% EC	Hiclear	Green	Slightly toxic	Caution	Botanical	Contact & Stomach	Antifeedant and repellent	5 ml	Whitefly, Bollworm, Thrips, Stem Borer, Brown Plant Hopper, Leaf folder	Cotton, Rice
Azadirachtin (Neem Seed Kernel Based)	0.30% EC	Margosom	Green	Slightly toxic	Caution	Botanical	Contact & Stomach	Antifeedant and repellent	4 ml	American boll worm	Cotton
Azadirachtin (Neem Based)	1 % EC	Neemarin, Ultineem	Green	Slightly toxic	Caution	Botanical	Contact & Stomach	Antifeedant and repellent	2-3 ml	Thrips, Red Spider mite, Fruit borer, Fruit and shoot borer	Tea, Tomato, Brinjal
Azadirachtin (Neem Oil Based)	0.03% EC	Super Killer	Green	Slightly toxic	Caution	Botanical	Contact & Stomach	Antifeedant and repellent	3-4 ml	Bollworm, Stem Borer, Brown Plant Hopper, Leaf roller, Aphid	Cotton, Rice
Azadirachtin (Neem Oil Based)	00.03% WSP	Super Killer	Green	Slightly toxic	Caution	Botanical	Contact & Stomach	Antifeedant and repellent	5 ml	Pod borer, Aphid, Jassid, Whitefly, Fruit borer, Fruit and Shoot borer, Beetles, Diamond back moth, Cabbage worm, Cabbage looper, Semi looper, Hairy caterpillar	Red gram, Cotton, Okra, Brinjal, Cabbage, Jute

Name	Formulation	Trade name	Colour	Toxicity	Label	Group	Mode	Action	Dose	Pests	Crops
Azadirachtin (Neem Extract Concentrates)	5 % EC	Neemazol F	Green	Slightly toxic	Caution	Botanical	Contact & Stomach	Antifeedant and repellent	0.5 ml	Caterpillar, Pink mite, Red spider mites, Thrips, Tobacco caterpillar, Aphids, Brown plant hopper, Leaf folder, Stem borer, Whitefly, Leaf hoppers, Diamond back moth, Fruit borer	Tea, Tobacco, Rice, Cotton, Cauliflower, Okra, Tomato
Bacillus thuringiensis var. galleriae	1.3% FC	-	Green	Slightly toxic	Caution	Microbial	Stomach	Antifeedant and repellent	2 ml	Diamond back moth, Fruit borer, Leaf folder	Cabbage, Cauliflower, Tomato, Okra, Cotton, Rice, Chilli
Bacillus thuringiensis var kurstaki	1 % WP	Halt	Green	Slightly toxic	Caution	Microbial	Stomach	Antifeedant and repellent	2 gm	Diamond back moth, Fruit borer, Leaf folder	Cabbage, Cauliflower, Tomato, Okra, Cotton, Rice, Chilli
Bacillus thuringiensis var kurstaki	1 % SL	Dipel	Green	Slightly toxic	Caution	Microbial	Stomach	Stomach poison	2 ml	Diamond back moth, Fruit borer, Leaf folder	Cabbage, Cauliflower, Tomato, Okra, Cotton, Rice, Chilli
Bacillus thuringiensis serovar kurustaki	5.0% WP	-	Green	Slightly toxic	Caution	Microbial	Stomach	Stomach poison	0.5 gm	American bollworm, Spotted bollworm, Semilooper, Tobacco caterpillar, Diamond back moth, Stem borer, Leaf folder	Cotton, Okra, Cabbage, Tomato, Rice, Tobacco, Gram
Beauveria bassiana	1.15 % WP	Daman	Green	Slightly toxic	Caution	Microbial	Contact	-	5 gm	Bollworm, Leaf folder, Diamond back moth, Spotted bollworm	Cotton, Paddy, Cabbage, Okra
Benfuracarb	03 % GR	Tadaaki	Red	Extremely toxic	POISON, skull with cross-bones	Carbamate	Stomach	Nerve poison	33 kg/ha	Stem borer, Leaf folder, Brown plant hopper, Nematode	Rice, maize, sugar beet, rice, vegetables

Common name	Formulation	Trade name	Colour	Toxicity	Label	Group	Mode	Poison type	Dose	Pest	Crop
Benfuracarb	40 % EC	Bendiocarb	Red	Extremely toxic	POISON, skull with cross-bones	Carbamate	Stomach	Nerve poison	5 ml	Pod borer	Rice, maize, sugar beet, rice, vegetables
Benzpyrimoxan	10% SC	Orchestra	Blue	Nerve poison	Danger	IGR	Contact	Chitin Synthesis Inibitor	1.5-2 g	Brown Plant Hopper, White Backed Plant Hopper	Rice
Beta-cyfluthrin	02.45 % SC	Responser	Yellow	Highly toxic	Poison	SPs	Contact	Nerve poison	1-1.5 ml	Boll worm	Tobacco
Bifenazate	50 % WP	Acramite	Green	Slightly toxic	Caution	Carbazate	Contact	-	0.25 g	Mite	Tea
Bifenazate	22.60 % SC	Swal	Green	Slightly toxic	Caution	Carbazate	Contact	-	0.25 ml	Mite	Tea
Bifenthrin	08 % SC	Chemet Nobu	Yellow	Highly toxic	Poison	SPs	Contact	Nerve poison	1.25 ml	Red spider mite, Tea Mosquito bug	Tea
Bifenthrin	08.80 % CS	-	Yellow	Highly toxic	Poison	SPs	Contact	Nerve poison	1 ml	Stem borer, Leaf folder	Rice
Bifenthrin	10 % EC	Histar	Yellow	Highly toxic	Poison	SPs	Contact	Nerve poison	1.6 ml	Bollworms, Whitefly, Stem borer, Leaf folder, Green leaf hopper, termites	Cotton, Rice
Brodifacoum	0.005 %w/w BB	Arakus, Jaguar, Rattex, Rodend, Rodenthor, Ratsak, Talon, Volak, Vertox, and Volid	Red	Extremely toxic	POISON, skull with cross-bones	Anticoagulent	Stomach	Nerve poison	One bait of 0.005% (a block of 20 gm each) per baiting station as a single feed	Field rats/Bandicoot rats (*Bandeicota bengalensis*; *B. indica*) Indian house rat / Black, Indian house rat/black rat/roof rat (*Rattus rattus*; *R. meltade*),	-

Insecticide	Formulation	Trade name	Colour	Toxicity	Caution	Chemical class	Mode of action	Poison	Dose	Pest	Crop
Broflanilide	300 g/l SC	Teraxxa	Green	Slightly toxic	Caution	meta-diamides and isoxazolines	Stomach and translaminar	Nerve poison	0.125 ml	Fruit borer (*Helicoverpa armigera*), Thrips (*Scirtothrips dorsalis*), Shoot and fruit borer (*Leucinodes orbonalis*), Diamond back moth (*Plutella xylostella*), tobacco leaf eating caterpillar (*Spodoptera litura*), leaf miner (*Liriomyza trifolii*), *Maruca vitrata*	Corn, Tomato, Chilli, Okra, Cotton, Cabbage, Pulses
Broflanilide	20% SC	Exponus	Green	Slightly toxic	Caution	meta-diamides and isoxazolines	Stomach and translaminar	Nerve poison	0.25 ml		
Bromadiolone	00.25 % CB	Ratcon	Red	Extremely toxic	Poison with skull	Anticoagulent	Systemic & Stomach	Nerve poison	0.005 %	Field Rat, Large Bandicota Indian house rat, Indian field mouse	-
Bromadiolone	00.005 % RB	Kalrat RB	Red	Extremely toxic	Poison with skull	Anticoagulent	Systemic & Stomach	Nerve poison	0.005 %		
Buprofezin	25 % SC	Applaud	Green	Slightly toxic	Caution	Chitin synthesis inhibitors	Stomach & Systemic	IGR	1.5 ml	Whitefly, Aphid, Jassid, Yellow mite, Mealy bug, Hopper, Thrips	Vegetables, Cotton, Rice, Mango
Buprofezin	70 % DF	Zenny	Green	Slightly toxic	Caution	Chitin synthesis inhibitors	Stomach & Systemic	IGR	0.5 ml	Jassid, Whitefly, Brown Plant Hopper	

Common name	Formulation	Trade name	Label colour	Toxicity	Symbol	Chemical group	Mode of action	Poison	Dose	Pests	Crops
Carbofuran	3 % CG	Furadon, Coradan	Red	Extremely toxic	Poison with skull	Organo carbamate	Systemic & Stomach	Nerve poison	12 kg/ acre	Aphid, Jassid, Shoot fly, cyst nematode, Stem borer, White grub, Nematode, Pod borer, Whitefly, Wooly aphid, Leaf miner, Thrips, brown plant hopper, Hispa, Gall midge, Top borer, Rhizome weevil, Scale insect, Stem weevil	Rice, Vegetables, Pulses, Cotton, Sugarcane, Banana
Carbosulfan	6 % G		Red	Extremely toxic	Poison with skull	Organo carbamate	Contact and stomach	Nerve poison	16.7 Kg/ha	Stem borer, Gall midge, Green leaf hopper, Leaf folder	Rice, Cotton, Tube rose, Vegetables
Carbosulfan	25% EC	Marshal, Atanka	Yellow	Highly toxic	Poison	Organo carbamate	Contact and stomach	Nerve poison	2 ml	Green leaf hopper, White Back Plant Hopper, Brown plant hopper, Gall midge, Stem borer, leaf folder, Aphid, Thrips, Fruit and shoot borer	
Carbosulfan	25 % DS	Marshall	Red	Extremely toxic	Poison with skull	Organo carbamate	Contact and stomach	Nerve poison	60 g/Kg seed	Jassid, Aphids, Thrips	
Cartap Hydrochloride	04 % Granules	Padan, Critap	Yellow	Highly toxic	Poison	Organo carbamate	Contact and stomach	Nerve poison	20 Kg/ha	Stem borer, Leaf folder, Whorl maggot	Rice, Vegetables, Sugarcane, Fruits
Cartap hydrocloride	50 % SP	Padan, Kaldan	Yellow	Highly toxic	Poison	Organo carbamate	Contact and stomach	Nerve poison	1 gm	Stem borer, Leaf folder	
Cartap Hydrochloride	75 % SG	Tityus	Yellow	Highly toxic	Poison	Organo carbamate	Contact and stomach	Nerve poison	1 gm	Yellow stem borer, Leaf folder	

Insecticide	Formulation	Trade name	Colour	Toxicity	Caution/Poison	Group	Mode of entry	Mode of action	Dose	Target pests	Crops
Chlorantraniliprole	18.5 SC	Coragen	Green	Slightly toxic	Caution	Diamide	Stomach	Ryanodine muscle inhibitor	0.3 ml	Stem borer, Leaf folder, American bollworm, Spotted bollworm, Tobacco caterpillar, Dimond back moth, Termite, Early shoot borer, Fruit borer, Top borer, Fruit and Shoot borer, pod borer, Semilooper, Stem fly, Girdle beetle,	Rice, Brinjal, Tomato, Chilli, Cabbage, Cauliflower, Okra, Cotton, Pulses, Sugarcane
Chlorantraniliprole	0.4 G	Fertera	Green	Slightly toxic	Caution	Diamide	Stomach	Ryanodine muscle inhibitor	10-18.75 Kg/ha	Yellow stem borer, Leaf folder, Early shoot borer, Top borer	Rice, Brinjal, Tomato, Chilli, Cabbage, Cauliflower, Okra, Cotton, Pulses, Sugarcane
Chlorantraniliprole	35 % WG	-	Green	Slightly toxic	Caution	Diamide	Stomach	Ryanodine muscle inhibitor	0.18 g	Fruit borer (Helicoverpa armigera & Earias vittella)	Rice, Brinjal, Tomato, Chilli, Cabbage, Cauliflower, Okra, Cotton, Pulses, Sugarcane
Chlorofenapyr	10 % SC	Intrepid	Yellow	Highly toxic	Poison	Pyrroles	Contact, & Systemic	disrupting the production of adenosine triphosphate	2 ml	Dimond back moth and mites	Cabbage, Brinjal
Chlorfluazuron	05.40 % EC	Araboron	Green	Slightly toxic	Caution	IGRs	Contact and stmach	Chitin Synthesis Inhibitor	3 ml	Diamond back moth, Tobacco leaf eating caterpillar, American boll worm	Cabbage, Cauliflower, Tomato, Okra

Name	Formulation	Brand	Colour	Toxicity	Poison label	Chemical group	Mode	Poison type	Dose	Pests	Crops
Chlorpyriphos	20 % EC	Dursban, Dermet	Yellow	Highly toxic	Poison	Organo phosphate	Contact & Stomach	Nerve poison	2.5 ml	Gall midge, Stem borer, Whorl Maggot, Hispa, Pod borer, Black bug, Cut worm, Early shoot and top borer, Pyrilla, Aphid, Whitefly, Boll worm, Root grub, Shoot and fruit borer, diamond back moth, Ground beetle, Leaf hopper, Termite	Rice, Pulses, Sugarcane, Wheat, Vegetables
Chlorpyrifos	10 % Granules	Shibalik	Yellow	Highly toxic	Poison	Organo phosphate	Contact & Stomach	Nerve poison	10 Kg/ha	Stem borer, Leaf folder, Gall midge	Rice
Chlorpyrifos	50 % EC	Terminator	Yellow	Highly toxic	Poison	Organo phosphate	Contact & Stomach	Nerve poison	2 ml	Stem borer, Leaf folder, Boll worm	Rice, Cotton
Chlorpyriphos	75 % w/w WG	-	Yellow	Highly toxic	Poison	Organo phosphate	Contact & Stomach	Nerve poison	1 g	Yellow stem borer (Scirpophaga 12incertulas)	Rice
Chlorpyrifos	01.50 % DP	Pyrum	Yellow	Highly toxic	Poison	Organo phosphate	Contact & Stomach	Nerve poison	25 Kg/ha	Stem borer, Green leaf hopper, Brown plant hopper, Leaf folder, Gall midge, Grass hopper, Pod borer	Rice, Pulses
Chromafenozide	80 % WP	Dodger	Green	Slightly toxic	Caution	IGR	Contach and Stomach	Chitin synthesis inhibitor	1 gm	Leaf folder and Stem borer	Rice
Clothianidin	0.5 % GR	-	Yellow	Highly toxic	Poison	Neonicotinoids	Systemic	Nerve poison	10 Kg/ha	Brown Plant Hopper, jassid and Whitefly	Rice, Vegetables
Clothianidin	50 % WDG	Dantp, Dontorsu	Yellow	Highly toxic	Poison	Neonicotinoids	Systemic	Nerve poison	0.1 g	Brown Plant Hopper, Jassid, Whitefly, Thrips, Termite, Mosquito bug, Aphid	

Name	Formulation	Trade name	Colour	Toxicity	Signal word	Chemical class	Mode of entry	Mode of action	Dose	Pest	Crop
Coumatetralyl	0.75 % w/w Gel	-	Red	Extremely toxic	Poison with skull	Anticoagulent	Stomach	Nerve poison	2.50 mg per spot	Rats (Rattus rattus, Rattus norvegicus, Bandicota bengalensis, Bandicota indica	Stored and field grains
Coumatetralyl	0.0375 % Bait	-	Red	Extremely toxic	Poison with skull	Anticoagulent	Stomach	Nerve poison	2.50 mg per spot	Rats (Rattus rattus, Rattus norvegicus, Bandicota bengalensis, Bandicota indica, Terra indica, Meriones hurrianae), mice	Stored and field grains
Cyantraniliprole	10.26 % OD	Biostar, Benevia	Green	Slightly toxic	Caution	Anthranilic diamide	Systemic, Contact and Stomach	Uncontrolled release and depletion of calcium from muscle cells, thus preventing further muscle contraction and ultimately leading to death	1.2 ml	Thrips, Flea beetle, Whitefly, Cabbage aphid, Fruit borer, Tobacco caterpillar, Diamond back moth, Leaf miner, Whitefly, Red pumpkin beetle, Pumpkin caterpillar	Grapes, Pomegranate, Cabbage, Chilli, Tomato, Gherkins
Cyenopyrafen	30 % SC	Kunoichi	Green	Slightly toxic	Caution	Pyrazole	Contact	inhibitors of mitochondrial complex II	0.25-0.5 ml	Mite	Apple, Chilli
Cyflumetofen	20 % SC	Foster, Danisaraba	Blue	Moderately toxic	Danger	Benzoyl acetonitrile	Contact	inhibit mitochondria complex II	1.5 ml	Red spider mite	Tea
Cypermethrin	25 % DP	Cyper-Maxx	Yellow	Highly toxic	Poison	Synthetic pyrethroid	Contact & Stomach	Nerve poison	20 Kg/ha	Fruit and Shoot borer	Brinjal
Cypermethrin	10 EC	SRL	Yellow	Highly toxic	Poison	Synthetic pyrethroid	Contact & Stomach	Nerve poison	1 ml	Spotted bollworm, American bollworm, Pink bollworm, Dimond back moth, fruit borer, Fruit and Shoot borer, Shoot fly, Bihar hairy caterpillar	Cotton, Okra, Brinjal, Sunflower, Wheat

Insecticide	Formulation	Trade name	Colour	Toxicity	Signal word	Chemical group	Action	Mode of poison	Dose	Target pests	Crops
Cypermethrin	25 EC	Auzar	Yellow	Highly toxic	Poison	Synthetic pyrethroid	Contact & Stomach	Nerve poison	0.5 ml	Bollworm, Jassid, Thrips, Shoot and Fruit borer, Hudda beetle	Cotton, Brinjal, Okra
Deltamethrin	11% w/w EC	Decis	Yellow	Highly toxic	Poison	SPs	Contact	Nerve poison	0.4 ml	Stem borer, Leaf folder, Green leaf hopper, Whorl maggot, Thrips, Leaf folder, fruit borer	Cotton, Rice, Tea, Tomato, Okra, Chilli, onion
Deltamethrin	1.8 EC	Delrox	Yellow	Highly toxic	Poison	SPs	Contact	Nerve poison	1.25 ml	Bollworm, Sucking insects, Stem Borer, Leaf folder	Cotton, Rice
Deltamethrin	02.50 % WP	Savier	Yellow	Highly toxic	Poison	SPs	Contact	Nerve poison	2 g	Rice weevil, Leaser grain borer, Khapra beetle, Red flour beetle, Saw toothed grain beetle, Rice moth, Almond moth, Rice weevil, Leaser grain borer	Wheat, Rice seed grains, Walls and ceilings
Deltemethrin	2.8 % EC	Decis	Yellow	Highly toxic	Poison	Synthetic pyrethroid	Contact & Stomach	Nerve poison	1 ml	Stem borer, Leaf folder, Green leaf hopper, Whorl maggot, Thrips, Leaf folder, fruit borer, Leaf miner	Cotton, Rice, Tea, Tomato, Okra, Chilli, Onion, Groundnu, Brinjalt
Dicofol	18.5 % EC	Kelthane, Hilfol	Blue	Moderately toxic	Danger	Organo chlorinated	Contact	Nerve poison	2 ml	Red spider mite, Scarlet mite, Pink mite, Purple mite, Yellow mite	Tea, Okra, Citrus, Litchi, Cotton, Brinjal, Bottle gourd
Diafenthiuron	47.80 % SC	Dialog, Poloride	Yellow	Highly toxic	Poison	Thiourea derivative	Contact and stomach	inhibits mitochondrial functioning	1 ml	Whiteflies, Aphids, Thrips, Jassids	Cotton
Diafenthiuron	50 % WP	Peagasus	Yellow	Highly toxic	Poison	Thiourea derivative	Contact & Systemic	inhibits mitochondrial functioning	1 gm	Whiteflies, Aphids, Thrips, Jassids, Diamond back moth, mite, capsule borer	Cotton, Cabbage, Chilli, Brinjal, Citrus, Watermelon, Okra, Tomato
Diflubenzuron	25 % WP	Larvak	Green	Slightly toxic	Caution	IGR	Contact and stomach	Chitin synthesis inhibitor	0.6 ml	Bollworm, Tobacco caterpillar	Cotton

Diflubenzuron	25 % WP	Dimilin	Green	Slightly toxic	Caution	IGR	Contact & Stomach	IGR	0.5 ml	Diamond back moth, Bollworm, Tobacco caterpillar	Cabbage, Cotton
Dimethoate	30 % EC	Rogor, Tara 909	Yellow	Highly toxic	Poison	Organo phosphate	Contact, Systemic & Stomach	Nerve poison	2 ml	Aphids, Thrips, Jassids, Diamond back moth, mite, capsule borer, Stem borer, Aphid, Jassid, Thrips, Mealy bug, Scale insect, Shoot and fruit borer, Great weevil, Leaf miner, Saw fly	Mustard, Cabbage, Cauliflower, Chilli, Onion, Potato, Banana, Maize
Dichlorvos (DDVP)	76 % EC	Nuvan, Vapona, Suchlor	Yellow	Highly toxic	Poison	Organo phosphate	Contact, Stomach and fumigant	Nerve poison	0.75 ml	Aphid, Spider mite, Whitefly, Fruit fly, Thrips	Vegetables, Fruits
Dinotefuran	20 SG	Token	Yellow	Highly toxic	Poison	Neonicotinoids	Systemic	Nerve poison	0.3 g	Brown plant hopper, Green leaf hopper, White fly, Aphid, Thrips	Rice, Cotton
Emamectin benzoate	05 % SG	Proclaim	Yellow	Highly toxic	Poison	Avermectin	Stomach and translaminar activity	Inhibit muscle contraction	0.5 g	Bollworm, Shoot borer, Diamond back moth, Thrips, mites, Pod borer, Thrips	Cotton, Okra, Cabbage, Chilli, Brinjal, Red gram, Chickpea, Tea, Grapes
Emamectin benzoate	01.90 % EC	Larvi green	Yellow	Highly toxic	Poison	Avermectin	Stomach and translaminar activity	Inhibit muscle contraction	1 ml	Boll worm, Thrips, Pod borer, Leaf folder, Semilooper, Hispa,	Cotton, Chilli, Chick pes, Paddy
Ethion	50 EC	Umite, Rusmite, fosmite, Fosmite, Miticil	Yellow	Highly toxic	Poison	Organo phosphate	Contact & Stomach	Nerve poison	800 ml/ha	Red spider mites, Purple mites, Yellow mite, Thrips, Scale	Tea, Cotton, Chilli, Gram, Pigeon pea,

Name	Formulation	Trade name	Colour	Toxicity	Signal word	Chemical group	Action	Mode of action	Dose	Pests	Crops
Ethofenoprox	10 % EC	Trebon Excel	Blue	Moderately toxic	Danger	Aromatic ether	Contact	Nerve poison	1.5 ml	Brown plant hopper, Green leaf hopper, Stem borer, Leaf folder, Gall midge, Whorl maggot, White backed plant hopper	Rice
Etoxazole	10 % SC	Borneo Sumitomo	Blue	Moderately toxic	Danger	Pyrroles	Local systemic activity	Inhibition of the moulting process	0.8 ml	Red spider mite	Brinjal, Tea
Fenazaquin	10 % EC	Magister	Yellow	Highly toxic	Poison	Quinazoline	Contact and stomach	Disruption of the biochemistry of insect mitochondria.	2 ml	Red mites, Yellow mites, Purple mites, Pink mite	Tea, Chilli, Apple, Okra, Brinjal, Tomato
Fenitrothion	50 EC	Sigma, Sumithion	Yellow	Highly toxic	Poison	Organo phosphate	Contact	Nerve poison	1.5 ml	Sucking and biting insects	Rice, Stored grains
Fenthion	82.5 EC	Labaycid, Baycid	Yellow	Highly toxic	Poison	Organo phosphate	Contact & Systemic	Nerve poison	1 ml	Aphids, Mites, Fruit flies, Mango weevil	
Fenobucarb (BPMC)	50 % EC	Wardan	Blue	Moderately toxic	Danger	Organo carbamate	Contact	Nerve poison	2 ml	Brown plant hopper, Green leaf hopper	Rice
Fenpropathrin	10 % EC	Rody, Danitol, Meothrin, Herald, and Fenprodate	Yellow	Highly toxic	Poison	SPs	Contact	Nerve poison	1 ml	Pink boll worm, Spotted boll worm, American boll worm	Cottton
Fenpropathrin	30 % EC	Macothrin	Yellow	Highly toxic	Poison	SPs	Contact	Nerve poison	0.4 ml	Pink boll worm, Spotted boll worm, American boll worm, White fly, Thrips, Mites, Shoot and Fruit borer, Stem borer, Leaf folder	Cotton, Chilli, Brinjal, Okra, Tea, Paddy
Fenpyroximate	05 % EC	Sedna	Yellow	Highly toxic	Poison	Pyrazole	Contact and Stomach	Inhibit mitochondrial electron transport	0.75-1.25 ml	Red spider mite, Pink Mite, Purple mite, Yellow mite, Jassid, Eriophid mites	Tea, Chilli, Cotton, Coconut

Name	Formulation	Trade names	Colour	Toxicity	Poison	Group	Mode	Action	Dose	Target pests	Crops
Fenpyroximate	05 % SC	Neon	Yellow	Highly toxic	Poison	Pyrazole	Contact and Stomach	Inhibit mitochondrial electron transport	0.75-1.25	Red spider mite, Pink mite, Purple mite	Chilli, Tea
Fenvalerate	20 % EC	Superfen, Capfen, Tatafen	Yellow	Highly toxic	Poison	SPs	Contact and Stomach	Nerve poison	0.5 ml	Diamond back moth, American boll worm, Aphids, Jassids, Bollworm, Aphid, Jassid, Thrips, Shoot and fruit borer	Cabbage, cauliflower, Cotton, Okra, Brinjal
Fenvalerate	00.40 % DP	Fan Gold, Ghatak	Yellow	Highly toxic	Poison	SPs	Contact and Stomach	Nerve poison	25 Kg/ha	Spotted Bollworm, Pink Bollworm	Cotton
Fipronil	5 % EC	Regent, Argent	Yellow	Highly toxic	Poison	Pyrazoles	Contact, stomach and systemic	Nerve poison	1 ml	Stem borer, Brown plant hopper, Green leaf hopper, Rice leaf hopper, Rice Gall midge, Whorl maggot, White backed plant hopper, Diamond back moth, Thrips, Aphids, Fruit borers, Early shoot borer & Root borer, Aphid, Jassid, Thrips, White fly, Bollworm	Rice, Cabbage, Chilli, Sugarcane, Cotton
Fipronil	0.3 % G	Regent	Yellow	Highly toxic	Poison	Pyrazoles	Contact, stomach and systemic	Nerve poison	7.5 kg/acre	Stem borer, Brown plant hopper, Green leaf hopper, Rice gall midge, Whorl maggot, White backed plant hopper, Early shoot borer, Root borer, Termite	Rice, Sugarcane, Wheat

Name	Formulation	Trade names	Colour	Toxicity	Label	Group	Action mode	Mode of action	Dose	Pests	Crops
Fipronil	02.92 % EC	Printed, Rippen, Race	Yellow	Highly toxic	Poison	Pyrazoles	Contact, stomach and systemic	Nerve poison	2 ml	Ants, Beetles, Cockroaches, Fleas, Ticks, Termites, Mole Crickets, Thrips, Rootworms, Weevils	Cereals, Vegetables, Fruits, Household condition
Fipronil	18.87 % w/w SC	Cripton, Regent Gold, Faster Gold	Yellow	Highly toxic	Poison	Pyrazoles	Contact, stomach and systemic	Nerve poison	0.5 ml	Thrips, Thrips, Aphids, Helicoverpa armigera, Stem Borer, Leaf Folder, Brown Plant Hopper	Cotton, Chilli, Onion, Grapes
Fipronil	80 % WG	Jumbo, Jump, Hooter	Yellow	Highly toxic	Poison	Pyrazoles	Contact, stomach and systemic	Nerve poison	60 g/ha	Stem borer, Leaf folder, Thrips, Diamond back moth	Rice, Chilli, Cabbage
Flonicamid	50 WG	Ulala	Yellow	Highly toxic	Poison	pyridinecarboxamide	Systemic	Feeding behavior inhibited	0.3 g	Brown plant hopper, White backed plant hopper, Green leaf hopper, Aphids, Jassids, Thrips & Whiteflies	Rice, Cotton
Flubendiamide	20 % WDG	Takumi	Green	Slightly toxic	Caution	Diamide	Stomach & Translaminar activity	Disrupts proper muscle function	0.2 – 0.3 gm	Diamond back moth, Stem borer, leaf folder, American bollworm, Pod borer, Semilooper, Tobacco caterpillar, Early shoot borer	Cabbage, Cotton, Rice, Chilli, Tomato, Tea, Arhar, Black Gram, Green gram, Groundnut, sugarcane
Flubendiamide	39.35 SC	Fame	Green	Slightly toxic	Caution	Diamide	Stomach & Translaminar activity	Disrupts proper muscle function	0.2 ml	Diamond back moth, Stem borer, leaf folder, American bollworm, Pod borer, Semilooper, Tobacco caterpillar, Early shoot borer, Brinjal fruit and shoot borer	Cabbage, Cotton, Rice, Chilli, Tomato, Tea, Arhar, Black Gram, Green gram, Groundnut, sugarcane, Brinjal

Name	Formulation	Trade name	Colour	Toxicity	Signal	Chemical class	Mode of entry	Mode of action	Dose	Pest	Crop
Flubendiamide	00.70 % GR	Zygant	Green	Slightly toxic	Caution	Diamide	Stomach & Translaminar activity	Disrupts proper muscle function	14 Kg/ha	Stem borer	Rice
Flufenoxuron	10 % DC	Cascade	Green	Slightly toxic	Caution	IGR	Contact & Stomach	IGR	1 ml	Mite	Rose
Flufenzine	20 % SC	Flumite	Green	Slightly toxic	Caution	-	Stomach and Translaminar property	Inhibits the hatching of eggs.	1 ml	Red mite, Pink mite, Purple mite	Brinjal, Tea
Fluopyram	34.48 % w/w SC	Velum prime	-	-	-	Pyridylethylamide	-	selectively inhibits Complex II of the mitochondrial respiratory chain of nematodes	0.6 ml	Root knot nematode (*Meloidogyne incognita*)	Tomato
Flupyradifurone	17.09 % w/w SL	Sivanto	Blue	Moderately toxic	Danger	neonicotinoid	Systemic	Agonist of the insect nicotinic acetylcholine receptor (nAChR)	0.3 ml	Jassid, Whitefly	Okra
Flupyrimin	2% GR	Karmax	-	-	-	-	Systemic	Nicotinic antagonist	7.5 Kg/ha	Stem Borer, Brown Plant Hopper	Rice
Fluvalinate	25 % EC	Apistan, Klartan, Mavrik, Mavrik Aqua Flow, Spur, Taufluvalinate, and Yardex	Blue	Moderately toxic	Danger	SPs	Contact	Nerve poison	0.4 ml	Aphids, Jassids, Red cotton bug, Boll worm	Cotton
Hexythiazox	05.45 % w/w EC	Maiden	Green	Slightly toxic	Caution	IGR	Contact, stomach with translaminar effect	Moulting inhibitor	1 ml	Scarlet mite, Red spider mite, Yellow mite	Tea, Chilli, Grapes, Rose, Brinjal, Okra, Apple

Name	Formulation	Trade names	Colour	Toxicity	Poison	Group	Action	Poison	Dose	Pests	Crops
Imidacloprid	17.8 % SL	Confidor, Tatamida	Yellow	Highly toxic	Poison	Neonicotinoids	Systemic	Nerve poison	0.2 ml	Jassids, Aphids, Thrips, White Fly, Brown plant hopper, White backed plant hopper, Leaf minor, Psylla, Flea beetle	Cotton, Rice, Okra, Cucumber, Tomato, Potato, Mango, Sugarcane, Tomato, Chilli, Brinjal, Citrus, Groundnut, Grapes
Imidacloprid	70 % WDG	Admire, Gaucho	Yellow	Highly toxic	Poison	Neonicotinoids	Systemic	Nerve poison	0.2 ml	Jassids, Aphids, Thrips, White fly, Brown plant hopper, White backed plant hopper	Cotton, Rice, Okra, Cucumber, Tomato, Potato
Imidacloprid	48 % FS	Imigro, Loyal FS	Yellow	Highly toxic	Poison	Neonicotinoids	Systemic	Nerve poison	0.9 Kg/ha	Aphids, Whitefly, Jassids, Thrips. Shoot fly, Termite	Cotton, Okra, Sunflower, Sorghum, Pearl millet, Soybean, Maize, Potato, Rice, Wheat
Imidacloprid	70 % WS	Noble, Finish	Yellow	Highly toxic	Poison	Neonicotinoids	Systemic	Nerve poison	1 Kg/ha	Aphids, Whitefly, Jassids, Thrips, Shoot fly, Painted bug, Termite, Saw fly	Cotton, Okra, Chilli, Sunflower, Sugarcane, Sorghum, Pearl millet, Mustard
Imidacloprid	30.50 % m/m SC	Winner, Imida 305	Yellow	Highly toxic	Poison	Neonicotinoids	Systemic	Nerve poison	0.2 ml	Aphid, Jassids, Thrips, Brown plant hopper, White backed plant hopper	Cotton, Rice, Chilli,
Imidacloprid	00.30 % GR	Ronaldo	Yellow	Highly toxic	Poison	Neonicotinoids	Systemic	Nerve poison	15 Kg/ha	Stem borer	Paddy
Indoxacarb	14.5 % SC	Abant	Yellow	Highly toxic	Poison	Oxadiazine	Contact, Stomach & systemic	Nerve poison	1 ml	Bollworm, Fruit borer, Pod borer complex, Diamond back moth	Cotton, Tomato, Pulses, Sunflower, Chilli, Cabbage
Indoxacarb	15.80 % EC	Indo gold plus	Yellow	Highly toxic	Poison	Oxadiazine	Contact, Stomach & systemic	Nerve poison	0.5 - 1 ml	Bollworm, Fruit borer, Pod borer complex, Diamond back moth, Leaf folder	Cotton, Tomato, Pulses, Sunflower, Chilli, Cabbage, Rice, Soybean

Name	Formulation	Trade names	Colour	Toxicity	Poison/Caution	Chemical group	Mode of action	Nature	Dose	Pests	Crops
Lambda–cyhalothrin	04.90 % CS	Robin, Legend	Yellow	Highly toxic	Poison	Synthetic pyrethroid	Contact & Stomach	Nerve poison	0.5 – 1 ml	Bollworm, Fruit borer, Pod borer complex, Diamond back moth, Leaf folder, Stem borer, Thrips, Shoot and Fruit borer, Flea beetle, Stem fly Semilooper, Shoot and capsule borer	Cotton, Paddy, Brinjal, Okra, tomato, Grapes, Chilli, Soybean, Pomegranate, Cardamom
Lambda cyhalothrin	2.5 % EC	Kunfu,	Yellow	Highly toxic	Poison	Synthetic pyrethroid	Contact & Stomach	Nerve poison	0.75 ml	Leaf folder, Stem borer, Green leaf hopper, Gall midge, Hispa, Thrips, Aphid, Jassid,	Rice, Cotton
Lambda cyhalothrin	5 % EC	Agent plus, Karate, Reeva	Yellow	Highly toxic	Poison	Synthetic pyrethroid	Contact & Stomach	Nerve poison	0.3 ml	Leaf folder, Stem borer, Green leaf hopper, Gall Midge, Rice hispa, Thrips, Bollworms, Jassids, Thrips, Shoot and Fruit borer, Mite, Pod borer, Pod fly, Leaf miner, Hopper	Rice, Cotton, Brinjal, Tomato, Chilli, Pigeon pea, Onion, Okra, Ground nut, Mango
Lufenuron	05.40 % EC	Luron	Green	Slightly toxic	Caution	Benzoyl urea	Stomach	Chitin synthesis inhibitor	1.2 ml	Diamond back moth, Pod borer, Fruit borer, Pod fly	Cabbage, Cauliflower, Pulses, Cotton, Chilli
Lufenuron	48 % SC	Lufeluron, Match, Signa	Green	Slightly toxic	Caution	IGR	Stomach	Chitin synthesis inhibitor	1.5 ml	Diamond back moth, Pod borer, Fruit borer, Pod fly	Cabbage, Cauliflower, Pulses, Cotton, Chilli

Name	Formulation	Trade name	Colour	Toxicity	Signal word	Chemical group	Mode of action	Poison type	Dose	Pests	Crops
Malathion	50 EC	Hilmala	Blue	Moderately toxic	Danger	Organo phosphate	Contact & Stomach	Nerve poison	2 ml	Hispa, Earhead midge, Pod borer, Leaf weevil, Jassid, Semilooper, Whitefly, Aphid, Spotted boll worm, mites, Head borer, Stem borer, Tobacco caterpillar, Scale, Wooly aphid, Mealy bug, Beetles, Hoppers	Rice, Sorghum, Pea, Soybean, Castor, Sunflower, Okra, Brinjal, Cabbage, Cauliflower, Turnip, Tomato, Radish, Apple, Mango, Grape
Metaflumizone	22 % SC	Imazalil	Green	Slightly toxic	Caution	a semicarbazone insecticide	-	Nerve poison	2 ml	Diamond back moth	Cabbage
Metaldehyde	2.5% DP	Snailkill	Blue	Moderately toxic	Danger	Tetroxocanes	Contact and stomach	Disrupting the mucus production ability	20 Kg/ha	Snails, Slugs, Giant, African snails	Citrus, Rubber, Paddy (Rice), Tea, Vegetables
Methomyl	40 SP	Lannate	Red	Extremely toxic	Poison with skull	Organo carbamate	Contact & Stomach	Nerve poison	2 gm	Tobacco caterpillar, American caterpillar, Thrips, Mealy bug	Cotton, Arhar, Tomato, Chilli, Groundnut, Grapes
Methoxyfenozide	21.8 % w/w SC	Maxium	Green	Slightly toxic	Caution	IGR	Stomach	promotes abnormal molting	1.8 ml	Leaf eating caterpillar, leaf miner, Pod borer, Early shoot borer	Groundnut, Sugarcane
Milbemectin	01 % EC	Floramite	Green	Slightly toxic	Caution	Milbemycins	Contact, stomach and tanslaminar activity	Nerve poison	0.5 ml	Two spotted Spider mite, Yellow mite, White mite	Rose, Chilli
Monocrotophos	15 % SG	-	Red	Extremely toxic	Poison with skull	Organophosphate	Contact, stomach and systemic	Nerve poison	1.3 ml	Aphids, Jassids, Thrips, Whiteflies	Cotton

Insecticide	Formulation	Trade name	Colour	Toxicity	Poison label	Chemical group	Mode of action	Action type	Dose	Pests	Crops
Monocrotophos	36 % SL	Monostar, Chemocron	Red	Extremely toxic	Poison with skull	Organophosphate	Contact, stomach and systemic	Nerve poison	2 ml	Brown plant hopper, Yellow stem borer, Green leaf hopper, Leaf folder, Shoot fly, Pod borer, Leaf miner, Plume moth, Shoot borer, Mealy bug, Scale insect, , Stalk borer, Boll worm, Aphid, Thrips, Grey weevil, Whitefly, Black aphid, Mite, Green bug, Thrips	Paddy, Maize, Black gram, Green gram, Pea, Sugarcane, Cotton, Citrus, Coconut, Coffee, Cardamon
Metarhizium anisopliae	1 % WP	Biomet, Kalichakra	Green	Slightly toxic	Caution	Microbial	Contact	Colonization of haemolymph	5 ml	Rhinoceros beetle	Coconut
Novaluron	10 % EC	Rimon	Green	Slightly toxic	Caution	IGR	Growth regulator	IGR	1 ml	American bollworm, Diamond back moth, Tobacco caterpillar, Pod borer	Cotton, Cabbage, Tomato, Chilli, Pulses
Nuclear polyhedrosis virus (Ha NPV)	1×10^8 PIBs/ml	Helicide	Green	Slightly toxic	Caution	Microbial	Stomach	Produce crystals in the fluids of the host	1 ml	*Helicoverpa armigera*	Pulses
Nuclear polyhedrosis virus (Sl NPV)	1×10^8 PIBs/ml	Spodocide	Green	Slightly toxic	Caution	Microbial	Stomach		1 ml	*Spodoptera litura*	Pulses
Oxydemeton methyl	25 % EC	Metasystox	Yellow	Highly toxic	Poison	Organo phosphate	Contact & Systemic	Nerve poison	2 ml	Aphid, Jassid, Thrips, Whitefly, Mealy bug, Scale insect, Leaf miner, Hopper, Tingid bug, Wooly aphid	Rice, Maize, Pulses, Chilli, Tomato, Cotton, Groundnut, Mustard, Okra, Sesame, Onion, Potato, Mango,Tobacco
Permethrin	25 EC	Perkill	Yellow	Highly toxic	Poison	Synthetic pyrethroid	Contact & Stomach	Nerve poison	1 ml	Boll worm	Cotton
Profenophos	50 % EC	Carina, Kuracron	Yellow	Highly toxic	Poison	Organo phosphate	Contact & Stomach	Nerve poison	1 ml	Jassids, Aphids, Thrips, Whiteflies, Bollworm. Semilooper	Cotton, Soyabean

Name	Formulation	Trade name	Colour	Toxicity	Signal word	Chemical group	Mode of action	Mechanism	Dose	Target pest	Crop
Phorate	10 % G	Thimet	Red	Extremely toxic	Poison with skull	Organo phosphate	Systemic	Nerve poison	4 kg/acre	Broad spectrum	Rice, Sugarcane
Phosphamidon	40 SL	Don, Hydon, Dimecron	Red	Extremely toxic	Poison with skull	Organo phosphate	Contact, Systemic & Stomach	Nerve poison	2 ml	Broad spectrum	
Propargite	57 % EC	Omite, Simba	Yellow	Highly toxic	Poison	Sulfite ester	Contact	Interruption of normal metabolism, respiration and electron transport functions	2 ml	Red spider mite, Pink mite, Purple mite, Scarlet mite, Yellow mite	Tea, Chilli, Brinjal
Pymetrozine	50 WG	Chess	Blue	Moderately toxic	Danger	Triazine	Systemic and stomach	Effects on neuroregulation or nerve-muscle	0.6 gm	Brown Plant Hopper	Rice
Pyrifluquinazon	20% WG	Rycar	Blue	Moderately toxic	Danger	-	Contact and stomach	Stop insect feeding (Behaviour modifier)	1 gm	Whitefly	Cotton
Pyridaben	20 % w/w WP	Pyrida	Blue	Moderately toxic	Danger	Pyridazinone	Contact	Mitochondrial complex I electron transport inhibitor	1 ml	Red spider mite, White fly	Tea, Cotton, Chilli
Pyridalyl	10 % EC	Sumipleo	Yellow	Highly toxic	Poison	-	Systemic	Inhibits cellular protein synthesis	1.5 ml	Boll worm, Fruit and shoot borer, Diamond back moth	Cotton, Okra, Cabbage
Quinalphos	25 % EC	Ekalux, Flush	Yellow	Highly toxic	Poison	Organo phosphate	Contact & Systemic	Nerve poison	2 ml	Brown plant hopper, Hispa, Leaf folder, Stem borer, Mite, Shoot fly, Aphid, Pod borer, Behar hairy caterpillar, Stem fly, Pod fly, Leaf weevil, Semilooper, Leaf roller, Yellow mite, Saw fly, Leaf webber, Thrips, Scale insect, Citrus butterfly	Rice, Sorghum, Wheat, Bengal gram, Black Gram, Soyabean, Jute, Ground nut, Okra, Mustard, Sesamum, Okra, Citrus, Tea

Name	Formulation	Trade name	Colour	Toxicity	Category	Chemical group	Mode	Mode of action	Dose	Target pest	Crop
Spiromesifens	22.9 SC	Oberon	Blue	Moderately toxic	Danger	Tetraonic acid derivatives	Systemic	Decreasing lipid synthesis by inhibiting acetyl CoA carboxylase	0.5-1 ml	White fly, Mite	Brinjal, Cotton, Chilli, Okra, Tea, Tomato, Apple
Spinosad	45 % SC	Tracer, Sprinter, Conserve	Blue	Moderately toxic	Danger	Spinosyn	Contact and stomach	to alter the function of nicotinic and GABA-gated ion channels	0.25 ml	Fruit and Shoot borer, American bollworm, Thrips,,	Brinjal, Cotton, Chilli, Red Gram
Spinetoram	11.70 % SC	Delegate, Summit	Green	Slightly toxic	Caution	Spinosyn	Contact and stomach	to alter the function of nicotinic and GABA-gated ion channels	0.4 ml	Thrips, Tobacco caterpillar, Spotted bollworm, Fruit borer	Cotton, Chilli, Soyabean
Spirotetramat	15.31 % w/w OD	Mowento	Yellow	Highly toxic	Poison	Tetramic acid	Systemic	Lipid biosynthesis inhibition	0.4 ml	Thrips, Whitefly, Aphid, Mite, Mealy bug,	Chilli, Okra, Grapes
Tetraniliprole	18.18 SC	Vayego	Blue	Highly toxic	Poison	Anthranilic diamide	Contact, stomach and systemic	Inhibit activity of ryanodine receptor	0.6 ml	Yellow stem borer, Leaf folder, Girdle beetle, Spodoptera spp. Semilooper	Rice, Soybean
Tetraniliprole	40.34% FS	-	Blue	Highly toxic	Poison	Anthranilic diamide	Contact, stomach and systemic	Inhibit activity of ryanodine receptor	10 Kg/ha	Stem borer and leaf folder, Stem borer	Rice, Maize
Triazophos	40 % EC	Hostathion, Triphos	Yellow	Highly toxic	Poison	Organo phosphate	Contact & Stomach	Nerve poison	2 ml	Catterpillar & Sucking pests	
Thiodicarb	75 % EC	Larvin	Red	Extremely toxic	Poison with skull	Organo carbamate	Contact & Stomach	Nerve poison	1 gm	Diamond back moth, Bollworms, Diamond back moth, Shoot and fruit borer, Pod borer	Cabbage, Cotton, Brinjal, Chilli, Black Gram, Arhar
Thiacloprid	21.7 SC	Calipso, Alanto	Yellow	Highly toxic	Poison	Neonicotinoids	Systemic	Nerve poison	0.3 ml	Aphid, Jassid, Thrips, Mosquito bug, Fruit and Shoot borer, White fly, Stem Borer, Girdle beetle	Cotton, Paddy, Chilli, Tea, Brinjal, Soybean, Apple

Name	Formulation	Trade name	Label colour	Toxicity	Signal word	Group	Action	Mode	Dose	Pests	Crops
Thiamethoxam	25 % WG	Ektara, Maxima, Evident	Blue	Highly toxic	Poison	Neonicotinoids	Systemic	Nerve poison	0.2 gm	Stem borer, Gall midge, Leaf folder, White backed plant hopper, Brown plant hopper, Green leaf hopper, Thrips, Aphid, Jassid, White fly, Mosquito bug	Rice, Cotton, Mustard, Okra, Tomato, Brinjal, Tea, Potato, Citrus, Tomato
Thiamethoxam	30 % FS	Polo gold	Blue	Highly toxic	Poison	Neonicotinoids	Systemic	Nerve poison	10 g/Kg seed	Aphid, Jassid, Shoot fly, Termite, Thrips	Cotton, Sorghum, Wheat, Soybean, Chilli, Okra, Maize, Sunflower
Thiamethoxam	70 % WS	Cruiser	Blue	Highly toxic	Poison	Neonicotinoids	Systemic	Nerve poison	3-6 g/Kg seed	Aphid, Jassid, Shoot fly, Termite, Thrips	Cotton, Sorghum, Wheat, Soybean, Chilli, Okra, Maize, Sunflower
Tolfenpyrad	15 % EC	Keefun	Yellow	Highly toxic	Poison	Pyrazole	Contact	Inhibit cellular respiration	2 ml	Diamond back moth, Aphids, Jassid, Whitefly, Thrips, Hopper	Cabbage, Okra, Cotton, Chilli, Onion, Mango
Verticillium lecani	1.15 WP (1x108 CFU/gm min)	Biolin	Green	Slightly toxic	Caution	Microbial	Contact	Invade cuticle and body cavity	5 ml	Whitefly, Mealy bug	Cotton, Citrus

Question - Answer

1. What is insecticide?

The substances used to kill insects which include ovicides and larvicides (used against insect eggs and larvae respectively) are called insecticides.

2. What is the most common insecticide?

The most commonly used insecticides are the organophosphates, pyrethroids and carbamates.

3. What do you mean by pesticide label?

The documents are on how to mix, apply, store and dispose of a pesticide/insecticide and the manufacturers' main way to give the user information about the product, are called labels.

4. What information should be on a pesticide label?

Most pesticide labels must include a signal word. The signal words Danger, Warning, or Caution - appear in large letters on the front panel of the pesticide label. They indicate the acute toxicity of the product to humans.

5. How is insecticide toxicity measured?

Acute toxicity is measured by LD_{50} and LC_{50} values. The smaller the LD_{50}, the more toxic the pesticide. Example: a pesticide with an LD_{50} of 5 mg/kg is 100 times more toxic than a pesticide with an LD_{50} of 500 mg/kg.

6. How do insecticides cause toxicity to humans?

Many insecticides can cause poisoning after being swallowed, inhaled, or absorbed through the skin. Symptoms may include eye tearing, coughing, heart problems, and breathing difficulties.

7. Name the 4 color symbols which are used to indicate the toxicity level of insecticides.

The four colour symbols used to indicate the level of toxicity in insecticides are: red (extremely toxic), yellow (highly toxic), blue (moderately toxic) and green (slightly toxic).

8. What is the threshold level of toxicity?

When a chemical causes a defined form of toxicity, the threshold is the maximum exposure when this toxicity does not occur. It is an operational parameter and is limited in its interpretation and applicability.

9. What is 'inhalation toxicity'?

The toxic effect when poisonous fumes of insecticide are inhaled (fumigants), is called inhalation toxicity.

10. What is difference between acute and chronic toxicity?

Acute toxicity: It refers to the ability to do systemic damage as a result of one-time exposure to relatively large amounts of the chemical. Toxic effect produced by a single dose of a toxicant, generally of short duration.
Chronic toxicity: Toxic effects produced by the accumulation of small amounts of the toxicant over a long period of time.

11. What happens if you breathe in insecticide?

Many insecticides can cause poisoning after being swallowed, inhaled, or absorbed through the skin. Symptoms may include eye tearing, coughing, heart problems, and breathing difficulties

12. What is the difference between LD_{50} and LC_{50}?

LD_{50} is the abbreviation used for the dose which kills 50% of the test population. LC_{50} is the abbreviation used for the exposure concentration of a toxic substance lethal to half of the test animals

13. What is a high LD_{50}?

These symbols relate the oral LD_{50} value (mg/kg) of a pesticide to its toxicity symbol. Danger Poison - LD_{50} less than 500 mg/kg indicate high toxicity. Warning Poison - LD_{50} 500 to 1,000 mg/kg indicates moderate toxicity.

14. What are the levels of toxicity?

The four toxicity categories, from one to four are:

- Toxicity category I is Highly toxic and Severely irritating,
- Toxicity category II is Moderately toxic and Moderately irritating,

• Toxicity category III is Slightly toxic and Slightly irritating,

• Toxicity category IV is Practically non-toxic and not an irritant.

15. Some of the red-label and yellow-label pesticides were banned in the state of ____following the ___protests of 2011.

Kerala, Endosulfan

16. How pesticides enter the body?

Before a pesticide can harm you, it must be taken into the body. Pesticides can enter the body orally (through the mouth and digestive system); dermally (through the skin); or by inhalation (through the nose and respiratory system).

17. Write about some basic characteristics of a label?

Labels are the principals or sometimes the only contact between the supplier and the user of the product. A good label should not be too complex or technical. It must be used correctly by the user.

18. What do you mean by IGR?

IGR or Insect Growth Regulator is a substance that inhibits the life cycle of an insect. These are typically used as insecticides.

19. What is toxicity?

The toxicity of a pesticide is its capacity or ability to cause injury or illness.

20. Mention the two types of toxicity.

Acute and chronic toxicity.

21. What are the four routes of exposure to toxicity?

The four routes of exposure are dermal (skin), inhalation (lungs), oral (mouth), and eyes.

22. Pyrethrins are derived from which plant?

Chrysanthemum flower

23. What are the dangerous substances present in commercial insecticide?

Carbamates, organoposphate and dichlorobenzene.

24. Mention some of the symptoms of insecticide toxicity in humans

Dizziness, blurred vision, headache, skin rashes etc.

25. How does an extremely toxic insecticide symbolize?

"Poison" is written on the label with a skull and cross-bone

26. " DANGER", which type of insecticide is depicted by this?

Moderately toxic insecticide.

27. Highly toxic insecticide is symbolised by which colour?

Bright yellow

Fill in the blanks

1. ……………….. are the largest group of insecticides.
Ans. Organophosphates

2. …………… and ………………….. are used for the evaluation of synthetic pyrethroids against insects.
Ans. KD_{50} and KT_{50}

3. ………………………….. unit is used to express the percent of active ingredient of the insecticide (% = 1/100) in case of median lethal concentration (or LC_{50}).
Ans. Parts per million (ppm)

4. Different methods used for bio-assay are …………, ……….. and ……………..
Ans. Direct Assays, Indirect Assays based upon quantitative responses, Indirect Assays based upon quantal responses

5. …………………. is the unit to calculate the oral and dermal toxicity of an insecticides on a test animal.
Ans: LD_{50}

6. LD_{50} is defined as the ……………. required to kill 50% of the population.
Ans: LD_{50}

7. LD_{50} by oral route for highly toxic is .... mg/kg.

Ans: 51 – 501

8. LD_{50} by dermal route for moderately toxic is .... mg/kg.

Ans: 2001-20000

9. LD_{50} by oral route for moderately toxic is .... mg/kg.

Ans: 501-5000

10. In case of insects, the LD_{50} is expressed in the unit of ....

Ans: Microgram/ gram

11. The concentration of insecticide in the external medium required to kill the 50% of the test population is termed as ....

Ans: LC_{50}

12. The colour of identification on band on the label in lower triangle of highly toxic insecticide is ....

Ans: Bright yellow

13. The oral value of a moderately toxic insecticide is about ....

Ans: 501- 5000 mg/kg

14. Insecticides rules was illustrated in the year ....

Ans: 1971

15. Labels are .... documents providing directions on how to mix , apply, store and dispose of a pesticide product.

Ans. legal

16. DANGER, WARNING or CAUTION are the .... words on the label of insecticide bottles

Ans: signal

17. Products with the DANGER signal word are the ....

Ans: most toxic

18. Products with the signal word .... are lower in toxicity.

Ans: CAUTION

19. The yellow label depicts .... insecticides.

Ans: highly toxic

20. A highly toxic insecticide is ....

Ans: endosulfan

21. Moderately toxic insecticide is labelled by ....

Ans: blue colour

22. Malathion is an insecticide which is ....

Ans: moderately toxic

23. Green label depicts .... insecticide.

Ans: slightly toxic

State true or False

1. Toxic effects produced by the accumulation of small amount of the toxicant over a long period of time is called acute toxicity

Ans. False

2. The minimum quantity of insecticide per unit weight of the organism required for killing 50% population of the test organism is called median lethal dose

Ans. True

3.   Permethrin is an example of synthetic pyrethroid

Ans. True

4. LD_{50} by oral route for moderately toxic is 1- 50 mg/kg.

Ans: False.

5. LD_{50} by oral route for highly toxic is 51-500 mg/kg.

Ans: True.

6. LD_{50} by dermal route for highly toxic is 2001-20000 mg/kg.

Ans: False.

7. LD_{50} by dermal route for slightly toxic is more than 20000 mg/kg.

Ans: True

MCQs

1. Median lethal dose by the oral route for extremely toxic insecticide –

    a. 1-50 mg/kg
    c. 1-100 mg/kg
    b. 51-500 mg/kg
    d. More than 5000 mg/kg
    Ans: a. 1-50 mg/kg

2. Symbols and word in upper triangle for highly toxic insecticide –

    a. Poison
    c. Caution
    b. Danger
    d. Poison, skull with cross bones
    Ans: a. Poison

3. Medial lethal dose by the dermal route for slightly toxic insecticide –

    a. 1- 200 mg/kg
    c. 2001- 20000 mg/kg
    b. 201 – 2000 mg/kg
    d. More than 20000 mg/kg
    Ans: d. More than 2000 mg/kg

4. Colour of identification band on label in lower triangle for moderately toxic insecticides –

    a. Bright red
    c. Bright blue
    b. Bright yellow
    d. Bright green
    Ans: c. Bright blue

5. Median lethal dose by dermal route for extremely toxic insecticide –

    a. 201- 2000 mg/kg
    c. More than 20000 mg/kg
    b. 1-200 mg/ kg
    d. 2001 – 20000 mg/kg
    Ans: b. 1-200 mg/kg

6. Which of the following information is usually found on a pesticides label?

    a) direction of use
    c) product name
    b) caution
    d) all of the above
    Ans – d) all of the above

7. Ovicides and larvicides are type of

    a) Insecticides
    c) rodenticides
    b) herbicides
    d) fungicides
    Ans. – a) insecticides

8. The insecticides which have residual and long-term activity is called as

   a) Contact insecticide   b) porous insecticides
   c) Systemic insecticide   d) penetrating insecticides
   Ans c) systemic insecticide

9. The insecticide which have no residual activity is called

   a) Systemic insecticide   b) contact insecticide
   c) Porous insecticides   d) penetrating insecticides
   Ans. b) contact insecticide

10. Carbaryl is a

   a) Highly toxic   b) Moderately toxic
   c) Slightly toxic   d) None of the above
   Ans. a) highly toxic

Chapter - 9

Acquaintance of
Insecticide Formulation

Introduction

Insecticides are substances used to kill insects. They include ovicides and larvicides used against insect eggs and larvae, respectively. Insecticides are used in agriculture, medicine, industry and by consumers. Insecticides are claimed to be a major factor behind the increase in the 20th-century's agricultural productivity. Nearly all insecticides have the potential to significantly alter ecosystems; many are toxic to humans and/ or animals; some become concentrated as they spread along the food chain.

Formulation

A pesticide formulation is a mixture of chemicals which effectively controls a pest. Formulating a pesticide helps to improve its storage, handling, safety, application, or effectiveness.

Formulation is the process of transforming an insecticidal chemical into a product which can be applied by practical methods to permit its effective to apply to target pests.

Active Ingredient

Active ingredients are the chemical in pesticide products or that portion of a formulation which possesses biological properties that act to control or repel the pests.

Factors

For making a formulation, some factors need to be considered which includes the chemical and physical properties of the active ingredient and the inert materials, the type of application equipments to be used, the nature of the target surface, and the marketing and transport aspects of pesticide usage. There are also certain things to be pondered in regards to compatibility of the inert gradient with active ingredients, compatibility with container and the physical properties of the final combined product. The formulated product must be evaluated to analyse the homogeneity of the product, particle size, storage constancy, retention in the target surface, wetting, penetration and translocation to various parts of the plants, the residual nature, nature of deposit efficacy and its impact on environment, etc.

Formulation consists of

» Active ingredient: Specific chemical in an insecticide that controls the pest.

» Carrier: Organic solvent or mineral clay.

» Surfactants, stabilizers and dyes.

Formulation = Pure form (active ingredient) + additives

Chemistry involved in formulation

1. Sorption (liquid a.i. on to solid surface)

Adsorption (a.i. on the surface of solid particles)

Absorption (a.i. enters into the pores of the solid)

2. Solution

» The solute (Solid/ Liquid) is dissolved in a liquid solvent.

» The components of true solution can't be mechanically separated.

» Does not require agitation.
Eg. Monocrotophos 36% SL

3. Suspension

» Finely divided solid a.i. dispersed in liquid.

» Don't form a true solution.

» Solid particles will be suspended in liquid and need agitation while spraying.

4. Emulsion

» One liquid is dispersed in other liquid.

» Form milky suspension when dissolved in water

» Requires agitation.

» The a.i. is dissolved in oil based solvent.

» Emulsifying agent helps to prevent the separation.

Classification of Pesticide Formulations

Solid formulations

Ready to use (Dusts, Granules, Pellets)

Concentrates (WP, SP, WDG)

Liquid Formulations (EC, SC, SL)

Types of formulation

There are many kinds of formulations available in the markets, including liquid, solid and gaseous form. A single Active ingredient may be available in different formulations. In agriculture, different types of formulations are frequently used for the application of insecticides employed in pest control programmes. The formulations are named as Dusts, Wettable powders (WPs), Emulsifiable Concentrates (ECs), Suspension Concentrates (SCs) or Flowables, Water Soluble Powder (SPs), Solutions, Granules, Water Dispersible Granules (WGs), Ultra-Low-Volume (ULV) formulations, Aerosols, Controlled Release (CR) formulations, Baits, etc. The non-insecticidal ingredients of formulations are Solvents, Diluents and Surfactants.

Dusts (D/DP)

1. Dusts are the simplest of all formulations as no water is required to spray. An insecticide dust is a dry formulation of a contact insecticide.

2. Dusts consist of active ingredient + inert Carrier/diluents + anti caking agent.

3. The toxicant in a dust formulation ranges from 0.1- 25 % and the rest consists of carriers.

4. The dust formulations have a particle size less than 100 microns.

5. Decrease in the particle size normally increases the toxicity of the dust formulation.

Examples: Carbaryl 5% DP, Carbaryl 10% DP, Malathion 5% DP, Quinalhpos 1.5% DP, Cypermethrin 0.25% DP, Endosulfan 3% DP.

Granules (G)

1. Dusts having larger (0.25 – 2.4 mm) particle sizes are called granular formulations

2. Granulated formulations consist of small pellets containing 2-10 % of the active ingredient. Granules consist of active ingredient + inert carrier/diluents.

3. An insecticide granule is a dry formulation of a generally systemic insecticide, but in some cases some contact insecticide also formulated as granules.

4. The active ingredients in granular formulations are low and may vary from 1-40 %.

5. Granular formulation is used mostly used against soil insects like nematodes etc.

Examples: carbaryl 4% G, carbofuran 3% G, carfosulfan 6% G, cartap hydrochloride 4% G, Chlorpyriphos 10% G, endosulfan 4% G, etc.

Wettable Powders (WPs) or Water Dispersible Powders (WDPs)

1. It refers to a substance that does not dissolve in water but remains suspended in it.

2. Wettable powders contain active ingredient + inert diluent/carrier + wetting agent/sufactant + anti caking agent.

3. The active ingredients may vary from 15% - 95%.

4. Wettable powders contain more active ingredient than dust.

5. Wettable powders should never be used without dilution. For application agitation is needed to get soluble powder into solution otherwise there will be certain amount of sedimentation might occur.

Examples: *Bacillus thuringiensis* 1% WP, *Beauveria bassiana* 1.15% WP, Cyfluthrin 10% WP, Deltamethrin 2.5% WP, Sulphur 80% WP, Thiodicarb 75% WP.

Soluble Powders (SPs)

1. Water-soluble powders contain a finely ground water soluble solid which dissolves readily upon the addition of water.

2. A water soluble powder contains active ingredient + inert carrier/diluent + wetting agent/surfactant + anti- caking agent.

3. SPs are usually 50 % or more active ingredients and always require dilution.

4. Unlike wettable powders, they do not require constant agitation and do not form precipitate. Example: cartap hydrochloride 50% SP, methomyl 40% SP and acephate 75% SP etc.

Water Dispersible Granules (WGs)

1. Water dispersible granules formulation appears as small granules.

2. Water dispersible granules formulations are mixed with water for application, they break apart similar to wettable powder and constant agitation is needed to get soluble powder into solution.

3. The active ingredients become distributed throughout the spray mixture.

Emulsifiable Concentrates (EC)

1. The formulation is usually in liquid form and it contains the active ingredient, organic solvents and emulsifying agents.

2. EC contains active ingredient + solvent + emulsifier + wetting agent/ surfactant (spreading/sticking agent added before spray).

3. When an emulsifiable concentrate is added to water, the emulsifier causes the oil to disperse uniformly throughout the water phase, giving an opaque or milky appearance when agitated.

Examples: Alphamethrin 10% EC, Azadirachtin 0.03%, 0.1%, 0.15%, %, 10%, 25% EC, Chlorpyriphos 20% EC

Soluble Liquids (SLs)

1. The formulation consists of active ingredient + water miscible solvent + surfactant.

2. These resemble emulsifiable concentrates in viscosity and color, but they do not form milky suspensions when diluted with water.

3. Addition of a surfactant provides wetting power. Examples: Monocrotophos 36% SL, Phosphamidon 40% SL, Imidacloprid 17.8% SL.

Suspension Concentrate (SC) or Flowable (F)

1. A suspension concentrate or liquid formulation combines many of the characteristics of emulsifiable concentrates and wettable powders.

2. It is a suspension that requires further dilution.

3. Active ingredient is insoluble either in water on in organic solvent.

4. Flowables may show sedimentation of the solid materials in storage and agitation by shaking makes the sediment redispersed.

Examples: Chlorfenapyr 10% SC, Flubendiamide 39.35 SC, Indoxacarb 14.5% SC, Spinosad 45% SC, Thiacloprid 21.7% SC, Spiromesifen 2% SC.

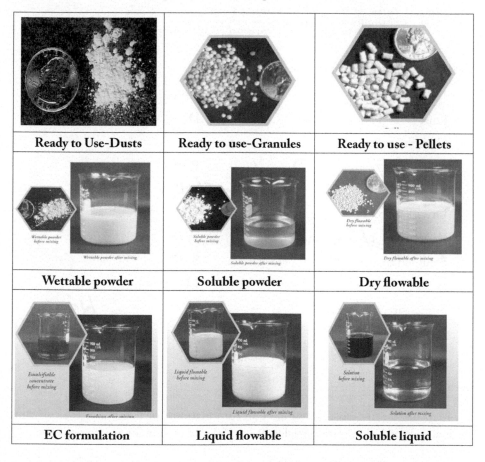

Ready to Use-Dusts	Ready to use-Granules	Ready to use - Pellets
Wettable powder	Soluble powder	Dry flowable
EC formulation	Liquid flowable	Soluble liquid

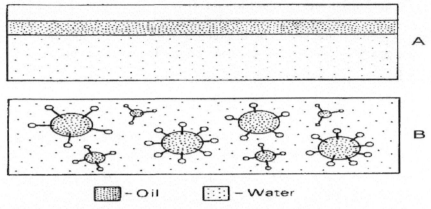

■ – Oil   ▦ – Water

(A)   Water and oil without emulsifier.
(B)   Emulsifiers link oil and water particles, enabling oil
       droplets to become suspended in water.

Short form of different formulations

Formulation	Suffix
Dust	D
Granule	G
Pellet	P, PS
Wettable Powder	WP, W
Dry Flowable	DF
Water Dispersible Granule	WDG
Soluble Powder	S, SP
Liquid Flowable	LF
Suspension Concentrate	SC
Microencapsulate	M
Emulsifiable Concentrate	EC

Aerosols

1. Aerosols are the most common of all formulations for home use. The toxicant is suspended as minute particles (0.1 to 50 microns) in air as a fog or mist.

2. To produce an aerosol, the active ingredients must be soluble in volatile petroleum oil under pressure.

3. When the solvent is atomized, it evaporates readily, leaving behind small droplets of the insecticide suspended in air.

4. Aerosols are used for the knockdown and control of flying insects and cockroaches, but they provide no residual effect.
Example: permethrin 0.2% and allethrin 0.1% are formulated as aerosols.

Fumigants

1. Chemical fumigants are gases or volatile liquids of low molecular weight which readily penetrate the material to be protected.

2. In most cases they are formulated as liquids under pressure and are stored in cans or tanks.

3. When released in open air, the liquid changes back into gas.

4. Fumigants are used for the control of insects in stored products but are also important for soil sterilization as a means of killing soil insects, nematodes, weed seeds, and fungi.

5. Example: Common fumigants used to treat stored products or nursery stock include hydrogen cyanide, naphthalene, nicotine, and methyl bromide.

Ultra-Low Volume Concentrates (ULV) or Concentrated Insecticide Liquids

1. These formulations are used on large areas of land to treat by spraying, i.e. hectares can be treated in one day.

2. These are applied with the help of droplet applicator (CDA).

3. They are applied without further dilution by aerial or ground equipment in volumes of 0.6 L to a maximum of 4.7 L /ha in very small spray droplets.

4. The advantage of ULV sprays is that the small droplets can better penetrate thick vegetation and other barriers.
Examples: Malathion, Dimethoate, Phosphamidon.

Poison baits

1. Formulated baits contain low levels of toxicants incorporated into materials such as food stuffs, sugar, molasses, etc. that are attractive to the target pest.

2.   The poison baits consist of a base or carrier material attractive to the pest species + a plant-based or chemical toxicant in relatively small quantities.

3.   The poison baits are used for the control of fruit flies, chewing insects, wireworms and white grubs in the soil, household pests (especially rodents) slugs etc.

Controlled Release (CR) formulations

These formulations which allow much less active ingredients to be used for the same period of activity. Controlled release formulations of pesticides are retarding repositories, which release their biologically active constituents into their environment over a defined period of time.

The non-insecticidal ingredients of formulations :

»   Solvents: A solvents is a substance, used for insecticides are soluble by dissolved. They are usually a liquid, but can also be a solid or gas.

    An example of solvents – a liquid, such as water, kerosene, xylene, or alcohol that will dissolve a pesticide to form a solution.

»   Diluents: Diluents are substance which combined with any concentrated insecticides and used to dilute an insecticide. Diluents can be either liquid or solid. Liquid diluents are usually water or refined oils while, solid diluents are used to formulate insecticide dusts or granules.

»   Surfactants: The term surfactants are derived from "Surface Active Agent". Surfactants may act as emulsifying, wetting agents, foaming agents and dispersants. Surfactants are materials that added to the insecticide formulation to enhance mixing.

Synergists

Synergists are chemicals which have little or no insecticidal property of their own, but when added to an insecticide, enhance its toxicity many fold. The enhanced toxicity is greater than the additive toxicity of the two taken together. This phenomenon is known as "synergism' and the substance producing it, a synergist. A substance that has the opposite effect is called an antagonist.

Question-Answer

MCQs

1. Thuriocide is

A) Insecticide   B) Weedicide   C) Fungicide   D) Antibiotic

Ans: A) Insecticide

2. Which of the following is an example of Dust formulation?

a. Methomyl   b. Carbaryl   C. Flubendiamide   d. Endosulfan

Ans: b. Carbaryl

3. Carbamates include

(A) Fungicides, herbicides and insecticides
(B) Insecticides only
(C) Herbicides, insecticides and nematicides
(D) Insecticides, nematicides an rodenticides
Ans: (A)Fungicides, herbicides and insecticides

4. The active ingredients in granular formulations vary from

a) 0.5-10%   b) 1-40%   c) 15-95%

Ans: b

5. Synergists –

a) Increases toxicity of insecticide
b) Decrease toxicity of insecticide
c) Acts as inert material

Ans: a

6. Example of Emulsifiable Concentrates is –

a) Methyl bromide   b) Azadirectin   c) Methomyl

Ans: b

7. What is the most common of all formulations for home use?

a) Aerosols   b) Fumigants   c) UVL

Ans: a

8. Surfactants acts as –

a) Emulsifying agent   b) Foaming agent   c) Both a & b

Ans: c

9. The purity of insecticide formulation is

a) 100%   b) 99%   c) < 80 %   d) 80-96 %

Ans: c

10. CMC is a

a) Maskers   b) Spreaders
c) Solvent   d) Wetting agents

Ans: d

11. EC is a type of

a) Solid   b) Liquid
c) Aerosols   d) Other insecticide formulation.

Ans: b

12. Chemicals that boost the activity of an insecticide is called

a) Synergists   b) Carrier   c) Solvent   d) Antagonists.

Ans. a

13. One dust additive is

a) Soap   b) Pine oil   c) Sulphur   d) Glue

Ans: c

14. Give an example of Water-Dispersible Granules (WDG) –

a) Acephate 75 SP   b) Carbaryl 50 WP
c) Indoxacarb   d) Permethrin

Ans. b

15. Granular formulations are dusts having particle size

a. 0.001-0.01 mm   b)10 mm   c) 1-2 mm   d) 0.25 – 2.4 mm

Ans: d

16. Which is a microbial insecticide?

a) *B. brevis*   b) *B. subtilis*   c) *Bacillus thuringenesis* d) *B. polymixa*

Ans: c

Fill in the blanks

1. Formulation = Pure form (active ingredient) + _____ .

Ans: additives

2. Permethrin 0.2% and _____ are formulated as aerosols.

Ans: allethrin 0.1%

3. Oil solutions are formulated by dissolving the insecticide in an _____ for direct use in insect control.

Ans: Organic solvent

4. Formulation =_____ + additives

Ans: Pure form (active ingredient)

5. Active ingredient is _____ that are responsible for the insecticidal effect.

Ans: Chemicals

6. Ultra-Low Volume Concentrates (ULV) usually contain a technical product dissolved in a minimum amount of _____

Ans: Solvent

7. ____ inert liquid or solid added to an active ingredient to prepare a insecticide formulation.

Ans: Carrier

8._____ are used for the knockdown and control of flying insects and cockroaches, but they provide no residual effect.

Ans: Aerosols

9. The dust formulations have a particle size less than ____microns.

Ans: 100 micron

10. Decrease in the particle size normally _____ the toxicity of the dust formulation.

Ans: increases

11. To produce an aerosol, the active ingredients must be soluble in _____ under pressure.

Ans: Volatile petroleum oil

12. _____ are substances used to kill insects.

Ans: Insecticides

13. An _____ is a mixture of chemicals which effectively controls an insect pest.

Ans: insecticide formulation

14. Dusts having larger (0.25 – 2.4 mm) particle sizes are called_____.

Ans: Granular

15._____are the most common of all formulations for home use.

Ans: Aerosols

SAQs

1. What are components of a pesticide formulation?

Pesticide Formulations: The pesticide active ingredient that controls the target pest. The carrier, such as an organic solvent or mineral clay. Adjuvants, such as stickers and spreaders. Other ingredients, such as stabilizers, softners, dyes, and chemicals that improve or enhance pesticidal activity.

2. What are the benefits of insecticide?

Humans have attained important benefits from many uses of insecticides, including: (1) increased yields of crops because of protection from defoliation and diseases; (2) prevention of much spoilage of stored foods; and (3) prevention of certain diseases, which conserves health and has saved the lives of millions of people and domestic animals. Pests destroy the potential yield of plant crops. Some of this damage can be reduced by the use of insecticides. In addition, insecticide spraying is one of the crucial tools used to reduce the abundance of mosquitoes and other insects that carry certain diseases (such as malaria) to humans. The use of insecticides to reduce the populations of these vectors has resulted in hundreds of millions of people being spared the deadly or debilitating effects of various diseases.

This is not to say that more insecticide use would yield even greater benefits. In fact, it has been argued that pesticide use could be decreased by one-half without causing much of a decrease in crop yields, while achieving important environmental benefits through fewer ecological damages.

3. What is the classification of Insecticides?

a. Solid formulations: Dust, wettable or water dispersible powder, granules, capsules, baits etc.

b. Liquid formulations: Solution, emulsifiable concentrate, ultra low volume formulations, suspension etc.

c. Gaseous formulations: Fumigant, aerosol, foams, smokes, mists and fog.

4. Will the formulation cause phytotoxicity (plant injury) to plants that are treated?

Phytotoxicity may occur following an insecticide application to plants. It is more likely to occur when using liquid formulations like EC formulations that contain solvents. When plants are to be treated,   select formulations that do not contain solvents (e.g., wettable powders, suspension concentrates, microencapsulated insecticides). Also, do not treat plants during the hottest parts of the day under direct sunlight. This combination of environmental factors can cause a plant's foliage to "burn."

Other causes of phytotoxicity may be attributed to using excessive rates, treating sensitive plant varieties, cumulative applications to the same plants and mixtures of products.

Manufacturers conduct extensive phytotoxicity trials with their products and can usually provide phytotoxicity information. If in doubt, first try the product on a small sample of plants.

5. Is the formulation effective against the target pest?

Regardless of the pest to be controlled, the objective must be to get the insecticide to the pest. Begin with proper pest identification. You should know if the pest is primarily a flying or crawling insect or if it spends most of its time in or below surfaces. Flying pests may be best controlled with a non-residual aerosol formulation. For crawling pests, a residual spray formulation or dust that stays on the surface where pests will contact it may be most appropriate. A bait formulation may be best, especially one that is attractive to the pest and placed in areas where they feed on the bait and ingest a lethal dose of the toxicant. Some active ingredients are selective in their activity against pests, so check the product label to be sure that the pest is listed.

6. Will the formulation perform well on the intended surfaces?

The interaction of residual insecticide formulations and the surface treated greatly affects insecticide performance. Residual insecticides are applied to a wide variety of surface types. These surfaces can be described as porous, semi-porous or non-porous types. Examples of porous surfaces include: concrete, wooden mulches, unfinished wood, gypsum wallboard, paper and plastics. Common nonporous surfaces are glass, ceramic tile and stainless steel. Semi-porous surfaces such as enamel- and latex-painted wood, vinyl tile and formica will also be encountered. Numerous studies have shown that EC formulations are absorbed into porous or semi-porous surfaces and the active ingredients are not available to crawling insect pests. Formulations like wettable powders, suspension concentrates, microencapsulated insecticides (CS) and dust formulations are not easily absorbed by porous surfaces and are more readily "picked up" by crawling insect pests. So, if the target pest is a crawling pest, select formulations that stay on the surface. However, if the target pest is found below surfaces (e.g., wood-boring beetles) an EC formulation is recommended. During the inspection and before application, determine which surface types are most common and identify the target pest(s). Once this information is gathered, you will be able to select the most effective formulation(s).

One of the biggest challenges facing the professional is getting cooperation from the customer to maintain good sanitation practices. In the real world of pest management, many of the surfaces treated are coated with oils, grease, food debris, dust or other organic matter, reducing the effectiveness of residual insecticides. Professionals can improve insecticide performance by encouraging their customers to maintain high levels of sanitation.

7. Is the product cost effective?

Several factors should be considered when answering this question. First, consider the use rate for the product and then determine the cost of treating a given area (e.g., cost per 1,000 square feet or per home or restaurant). Also, evaluate the time-saving features associated with the use of the product. Savings in labor costs can quickly offset higher insecticide costs. By using this product, can you reduce callback rates? The cost of callbacks is often underestimated. Often, the least expensive product is not the most cost effective .

8. Does the formulation mix easily and can it be tank mixed with other pesticides?

Other than ready-to-use products (e.g., baits, dusts, granules and aerosols), professionals must mix the products they use. Some formulations are easier to mix or handle than others. Liquids like EC, SC and CS formulations are usually preferred over dry formulations such as wettable powders (WP). However, the use of pre-measured, water soluble packaging for WP formulations has greatly increased the ease in handling this dry formulation. Check the label for compatibility with other pesticides. To be sure of physical compatibility, prepare a small sample of the mixture using the proper proportions of water and product.

9. Could there be visible residues on treated surfaces?

Unsightly visible residues on treated surfaces may occur following application of spray treatments. This problem is most common when using WP formulations at high labeled rates and spray volumes. The visible residue is primarily due to the carrier (e.g., talc, clay) used in the WP formulation. Visible residues appear to be more of a problem on indoor surfaces. When applying residuals indoors use EC, SC or CS formulations to minimize the risk of a visible residue.

11. What is a wettable powder formulation?

A wettable powder (WP) is a powder formulation that forms a suspension when mixed with water prior to spraying. WP formulations consist of one or more active ingredients which are blended and mixed with inerts, diluents and surfactants. Wetting agents are used to facilitate the suspension of the particles in water.

12. What is insecticide addictives?

The additives vary accordingly as the formulation is a solid (dust), liquid (spray) and gas (fumigant).

13. What do you mean by emulsifiers?

The emulsifier will allow the insecticide to form a stable emulsion of small globules on mixture with water in the spray tank. Thus, the insecticide emulsion can be sprayed easily onto the crop. Generally, the term formulation is reserved for commercial preparations of insecticide before they are sold.

14. Mention any 2 terms which are releted to the formulation chemistry of insecticide.

Sorption and Emulsion are two terms related to formulation of chemistry.

15. What do you mean by dust formulation?

Dust (D) formulations are dry with the active ingredient bound to clay or some other fine powder, or made entirely from pure active ingredient such as silica or borate. Wettable powder (WP) pesticides are dust-like formulations that are mixed with water, or sometimes oil, and sprayed through a sprayer.

16. Give 2 examples of Soluble Powder.

Cartap hydrochloride 50% SP, Methomyl 40% SP

17. Define formulation.

Formulation is the process of transforming an insecticidal chemical into a product which can be applied by practical methods to permit its effective to apply to target pests.

18. State a similarity and dissimilarity between soluble liquid and emulsifiable concentrate.

Soluble liquid resemble emulsifiable concentrates in viscosity and color, but they do not form milky suspensions (like EC) when diluted with water.

19. What are the non-insecticidal ingredients of formulations?

The non-insecticidal ingredients of formulation are solvents, dilutants, surfactants, dyes, stabilizers, etc.

20. Define chemical fumigants.

Chemical fumigants are gases or volatile liquids of low molecular weight which readily penetrate the material to be protected.

21. Give two examples of ULV.

Two examples of ULV are malathion and dimethoate.

22. Write an example of dust formulation.

An example of dust formulation is - Carbaryl 5% DP.

23. How did insecticide formulation form?

Insecticide formulation formed with the mixture of pure form of the insecticide (active ingredient) and the additives

24. Write down the types of spray additives.

Different types of spray additives are: solvents, wetting agents, emulsifires, spreaders, stickers, stabilizers, softeners, maskers.

25. Classify formulations according to the state of application.

Formulation is classified into three types on the basis of state of application. Those are – solid formulations, liquid formulations, others (aerosols, fumigants etc.).

26. What is the difference between synergists and antagonists?

Synergists are chemicals which have little or no insecticidal property of their own, but when added to an insecticide, enhance its toxicity many folds. Antagonists are the substances which have opposite effects of synergists.

State true and false

1. Cypermethrin is a example of Granule.
Ans: False.

2. Aerosols are the most common of all formulations for home use.

Ans: True.

3. Most of the organic insecticides are water insoluble, and therefore, are first dissolved in a solvent and then mixed with water for making a final spray is called Spreaders.

Ans: False

4. Granules (G) are quite similar to dust formulation but they are larger and break down more slowly.

Ans: True.

5. Solutions are active insecticide ingredients dissolve readily in a liquid solvent that may be used directly or required diluting.

Ans: True.

6. A substance that has the opposite effect of synergist is called an antagonist

Ans: True

7. Soluble Liquids (SL) consists of active ingredient + water miscible solvent + fumigant:

Ans: False

8. All insecticides are pesticide, but all pesticides are not insecticide.

Ans: True

9. The purity of technical (commercial) form of insecticide is greater than the insecticide formulation.

Ans: True

10. Stabilizers are dust additives.

Ans: False

11. Aerosols are the most uncommon of all formulations for home use.

Ans: False

12. Aluminium phosphide, Zinc phosphide are the examples of fumigants

Ans: True

Chapter - 10

Calculation of Doses/Concentration of Different Insecticides

Formula 1: $C_1V_1 = C_2V_2$ (This formula is used for Spray formulations like EC, SL, SC, WP, WDG, SP, WS, WDP, DS etc.)

C_1 = Concentration of commercial formulation (%)

V_1 = Volume or amount of insecticide formulation (ml/g)

C_2 = Desired concentration of spray fluid (%)

V_2 = Volume or amount of spray fluid required (ml)

Problems:

1. How much quantity of Chlorpyriphos 20 EC is required to spray @ 0.05% for control of YSB in paddy where the spray fluid recommended is 150 litre/ha?

Solution: We know, $C_1V_1 = C_2V_2$

Here,

C_1 = Concentration of commercial formulation (%) = 20

V_1 = Volume or amount of insecticide formulation (ml) = ?

C_2 = Desired concentration of spray fluid (%) = 0.05

V_2 = Volume or amount of spray fluid required (ml)/ ha = 150 litre = 150 x 1000 ml

So, V_1 = (0.05 x 150 x 1000) / 20 = 375 ml

Ans. 375 ml Chlorpyriphos 20 EC is required to spray @ 0.05% for control of YSB in paddy where the spray fluid recommended is 150 litre/ha.

2. **How much quantity of Indoxacarb 14.5 SC for one acre is required to spray @ 0.02% for control of *Helicoverpa armigera* in cotton where the spray fluid recommended is 200 litre/ha?**

Solution: We know, $C_1V_1 = C_2V_2$

Here, C_1 = Concentration of commercial formulation (%) = 14.5

V_1 = Volume or amount of insecticide formulation (ml) = ?

C_2 = Desired concentration of spray fluid (%) = 0.02

V_2 = Volume or amount of spray fluid required (ml)/ha = 200 litre = 200 x 1000 ml

       (For 1 acre it is 80 litre = 80000 ml)

       So, V_1 = (0.02 x 80000) / 14.5 = 110.34 ml

Ans. 110.34 ml Indoxacarb 14.5 SC for one acre is required to spray @ 0.02% for control of *Helicoverpa armigera* in cotton where the spray fluid recommended is 200 litre/ha?

3. **Calculate the percent concentration of insecticide in the spray fluid, where 250 ml Monocrotophos 36 SL is recommended in 125 litres/ha spray fluid to control aphids in cotton.**

Solution: We know, $C_1V_1 = C_2V_2$

Here, C_1 = Concentration of commercial formulation (%) = 36

V_1 = Volume or amount of insecticide formulation (ml) = 250

C_2 = Desired concentration of spray fluid (%) = ?

V_2 = Volume or amount of spray fluid required (ml)/ha = 125 litre = 125 x 1000 ml

So, C_2 = (36 x 250) / 125 x 1000 = 0.072 %

4. **How much spray fluid of 0.5% concentration can be prepared from 300 gm of malathion 25 WP?**

Solution: We know, $C_1V_1 = C_2V_2$

Here, C_1 = Concentration of commercial formulation (%) = 25

V_1 = Volume or amount of insecticide formulation (ml) = 300

C_{2} = Desired concentration of spray fluid (%) = 0.5

V_{2} = Volume or amount of spray fluid required (ml) =?

So, V_{2} = (25 x 300) / 0.5 = 15,000 ml or 15 litre

Formula 2: $C_{1}V_{1}$ = 100 RA (This formula is used to calculate the quantity or amount of toxicant for dusts and granular formulations)

1. **How much quantity of Phorate 10G is required to treat 7000 sq.m area of paddy field when the recommended rate is 0.5kg a.i./ha?**

Solution:

Given

C_{1} = Concentration of commercial formulation (%) = 10

V_{1} = Quantity of insecticide formulation (kg) = ?

R = Recommended rate of pesticide application (kg a.i./ha) = 0.5

A = Area to be treated in ha = 7000 m² = 0.7 ha

So, V_{1} = (100 x 0.5 x 0.7) / 10 = 3.5 kg

2. **Calculate the quantity of Fenvalerate 0.4%G dust applied @ 0.2 kg a.i. for three hectare area.**

Solution:

Given

C_{1} = Concentration of commercial formulation (%) = 0.4

V_{1} = Quantity of insecticide formulation (kg) = ?

R = Recommended rate of pesticide application (kg a.i./ha) = 0.2

A = Area to be treated in ha = 3 ha

So, V_{1} = (100 x 0.2 x 3) / 0.4 = 150 kg

3. **How much quantity of Carbofuran 3G is required to treat 2500 sq.m area of sorghum field to control stem borer when the recommended rate is 0.35kg a.i./ha?**

Solution:

Given

C_{1} = Concentration of commercial formulation (%) = 3

V_1 = Quantity of insecticide formulation (kg) = ?

R = Recommended rate of pesticide application (kg a.i./ha) = 0.35

A = Area to be treated in ha = 2500 m^2 = 0.25 ha

So, V_1 = (100 x 0.35 x 0.25) / 3 = 2.92 kg

Question and Answer

SAQs

1. A recommendation for aphids calls for using Malathion at 4 kg active ingredient/ hectare. How much Malathion 30% WP would be needed per hectare?

Solution:

Required Malathion 30 % WP (Kg/ha) of Malathion (y kg) = 30 % of y kg = 4 kg

So, y = 4/30 % kg = 13.33 kg

2. Determine the quantity of insecticide formulation and the quantity of water needed for a residual spraying programme to treat 1000 shelters of having an average surface area of 50 m^2. Insecticide formulation required is the deltamethrin 2.5 WP (Wettable Powder). The dose to be applied is 0.025 g of a.i. (active ingredient) per m^2.

Solution:

The amount of the water needed to treat 2000 shelters = (D) × (B) = 0.040 l/m^2 × 50 m^2 (1 shelter) = 2 litres hence, 2 litres × (A) = 2 litres × 1000 shelters = 2000 litres or 2 m^3 of water.

The quantity required of a.i. for one shelter when a dose of 0.025 g of a.i/m^2 is required = (B) × (C) = 50 m2 × 0.025 g = 1.25 g (E).

The amount of the insecticide formulation needed to treat one shelter= 2.5 WP = 2.5% = 2.5 g of a.i. for 100 g of insecticide formulation, what quantity of formulation represents 1.25 g of a.i.? 100 g ⇒ 2.5 g of a.i. x? ⇒ 1.25 g of a.i. then (100 × (E) ÷ 2.5 = (100 × 1.25 g) ÷ 2.5 = 50 g of insecticide formulation (F).

The total amount of formulation needed to treat 2000 shelters. Calculation (A) × (F) = 1000 shelters × 50 g of formulation = 50000 g = 50 kg.

The quantities needed for the spraying programme to treat 1000 shelters are: 2000 litres of water and 50 kg of deltamethrin 2.5 WP.

3. In order to spray 0.020 l per m^2 of solution containing a dose of 0.005g of a.i, what is the quantity of the insecticide concentration required in the solution, having deltamethrin 2.5 WP formulation?

Solution:

The quantity of a.i./m^2 to be sprayed, e.g. 0.005 g/m^2 (A); The speed of the spraying should be constant to spread 0.020 l (B) per m^2 containing 0.005 g of a.i.;

The formulation is deltamethrin 2.5 WP, that means 2.5% or 2.5 g of a.i. per 100 g of formulation.

The amount of a.i. per 1 litre of solution= 0.02 l of solution to be spray must contained 0.005 g of a.i. then, 1 litre ÷ (B) = 1 litre ÷ 0.020 l = 50 (C), hence, (A) × (C) = 0.005 g × 50= 0.25 g or 0.0025% of a.i. 0.25g or 0.0025% of the insecticide concentration must be mixed with 1 litre of water allowing the insecticide solution to be sprayed at a dose of 0.005 g of a.i. per 0.020 l and per m^2.

4. What is the quantity of deltamethrin 2.5 WP formulation required for 200 litres of insecticide solution wanted, at an insecticide concentration of 0.025 %.

Solution:

X = (A × B × D) ÷ C = (0.025 × 200 × 1) ÷ 2.5 = 2 Therefore 2 kg of insecticide formulation has to be mixed with 200 litres of water to obtain the solution at the concentration required.

5. What is the number of parts of water required for 1 part of the pirimiphos-methyl 75% EC formulation at a concentration of 2.5%.

Solution:

X = (A ÷ B) - 1 = (75 ÷ 2.5) - 1 = 29

Therefore 29 parts of water is required to 1 part of the insecticide formulation. i.e. 1 part insecticide: 29 parts water (1: 29) if 1 part insecticide = 2 litres, then 29 parts water = 58 litres of water which means 2 litres insecticide + 58 litres water = 60 of insecticide solution.

6. Calculate the quantity of insecticide and water needed to impregnate 100 mosquito nets where the height is 1.5 m, the length 2 m, and width 1 m. The overlap band measures 0.30 m of width. The required dose of a.i. is 0.2 g/m^2 using a permethrin 50 % EC formulation.

Solution:

Calculation of the surface area (Sm) of the mosquito net.

H = height = 1.5 m, L = length = 2 m, W = width = 1 m, Sm = (L × W) + (H × L × 2) + (H × W × 2) + (a × b) = (2 × 1) + (1.5 × 2 × 2) + (1.5 × 1 × 2) + (0.30 × 1.5) = 11.45 m², Calculation of the quantity of the insecticide formulation (M) required according the following formula: M = T, (C ´ 10) ml/m² of netting, T - target dose, this must be expressed in mg/m², therefore 0.2g of a.i./m² = 200 mg of a.i/m². C - insecticide concentration of the formulation, e.g. permethrin 50 EC formulation, C = 50. (Where the insecticide formulation concentration is given in g/litre, this should be divided by 10 to convert it to a percentage, e.g. 500g/l ÷ 10 = 50 EC = 50%).

Then M = T ÷ (C × 10) = 200 ÷ (50 × 10) = 0.4 ml/m² Therefore 0.4 ml/m² × 11.45 m² = 4.58 ml per mosquito net. 4.58 ml of permethrin 50 EC formulation are required for one mosquito net having a surface of 11.45 m² . This is to determine the absorption capacity of water of the mosquito net. This operation consists of: Dipping the mosquito net into a bucket containing 2 litres of water for 2-3 minutes; Taking out the net and wringing out the water over the bucket; Measuring the water left in the bucket, e.g. 1.470 l; Determining the quantity of water absorbed by the mosquito net, which is 2.000 l (initial input) - 1.470 l (water left) = 0.530 l.

The absorption capacity of one mosquito net is therefore 0.530 l or 530 ml. Quantities required for 100 mosquito nets: 100 × 0.52542 litre = 52.542 litres of water 100 × 0.00458 litre = 0.458 litres of insecticide formulation The absorption capacity of a mosquito net will depend upon the textile type, e.g. a cotton mosquito net may absorb 1.5 - 2 litres and a nylon one, 0.4-0.8 litre.

7. Determine the quantity of molluscicide required to treat the river with 70% niclosamide formulation and where 4 mg/l of concentration is required. The product has to remain in the water for 9 hours to eliminate the molluscs.

Solution:

Calculation of the cross section area of the stream (A)

The cross-sectional area of flow for the stream can be calculated using the trapezium rule. Now we calculate the area of flow between adjacent measurements of depth (D), and add these individual areas together to calculate the total area. The formula is A = W ´ (D1 + D2 + D3 + D4 + D5 + D6 + D7.........) W = distance between measurements of depth (must be kept constant) D1, D2.... = depth measurements.

The example assumes that the measurements have been done; the measures are expressed in metres. Therefore the cross section area is 2.48 m² (A)

A = W × (D1 + D2 + D3 + D4 + D5 + D6 + D7) A = 0.6 × (0.4 + 0.7 + 0.79 + 0.78 + 0.6 + 0.55 + 0.32) = 2.48 m²

Calculation of the surface flow velocity (Vs)

The following formula must be applied: Vs (m/second) = D (distance expressed in metre), T (second) This operation consists in measuring the period of time (T) that a float will take to travel a given distance (D) given. In this example D = 100 metres and it assumes that the time measured is 169 second. then Vs = D ÷ T = 100m ÷ 169 second = 0.591m/second. Any buoyant material may be used as a float such as a lemon, an empty bottle, a ball etc. Several measurements should be carried out and averaged.

Calculation of the stream velocity (Vm)

The following formula must be applied:

Vm (metres/second) = Vs (metres/second) ´ C (constant coefficient) C value is a constant. However this value may change under the stream characteristics. Generally the value given to C is 0.85 and 0.93 for rivers with sandy bed and depths of more than 3 metres, then Vm = Vs × C = 0.591 m/second × 0.85 = 0.502 m/second

Calculation of the discharge (Dq) of the streams

The following formula must be applied: Dq (m3/second) = Vm ´ A, then Dq = Vm × A = 0.502 m/second × 2.48 m² = 1.24 m³/second or 4481 m³/hour

The concentration of 4 mg/l of 70% of molluscicide formulation required for one hour at a flow of 1 m³/second = 1g of molluscicide is needed per 1m³/second means that 3.6 kg of product per hour is required to obtain the desired concentration, then 3.6 kg/h × 4 mg/l × [100 ÷ 70] = 20.6

Therefore 20.6 kg of product per hour is required. The quantity of 70% niclosamide molluscicide required per hour in the river with discharge of 1.24 m³/second and where the product has to remain in the water for 9 hours to eliminate the molluscs= [20.6 kg × 1.24 m³/second] x 9 =230kg

So, 230 kg of 70% niclosamide will be required for 9 hours.

8. How application rates of pesticides are expressed?

Application rates are usually expressed as amount of pesticide product but sometimes they may be expressed as pounds of active ingredient or actual toxicant. Actual toxicant and active ingredient are practically synonymous.

9. Why conversion of liquids to smaller quantities is easy?

Conversion of liquids to smaller quantities is relatively easy and precise because suitable equipment such as measuring spoons are readily available. While scales sensitive enough to handle small quantities of solid materials are available, it is often more practical to use volumetric measures.

10. Why some conversion tables are grossly inaccurate?

Various conversion tables have been prepared on the premise that there are 200 to 300 teaspoons (roughly 2-3 pints) per pound of solid pesticide product. These tables are grossly inaccurate because of the wide variation in bulk density among solid pesticide formulations. For instance, a pint of almost any insecticide wettable powder will weigh much less than a pint of fungicide that has a high metal content.

11. What are the different types of insecticides?

Depending on the chemical nature, the insecticides are classified into 4 groups, which are listed below:

- Organic insecticides
- Synthetic insecticides
- Inorganic insecticides
- Miscellaneous compounds

12. What is the function of insecticide?

Insecticides are chemicals used to control insects by killing them or preventing them from engaging in undesirable or destructive behaviours. They are classified based on their structure and mode of action

13. What happens if you don't water in insecticide?

Failure to water in the granules can lead to treatment failure because the chemical can not be released from the granule and will not be available to the pests

14. Will rain wash off insecticide?

One day of rain won't totally wipe out the treatment, but it does make the treatment the less effective. If more rain takes place, both farmers or home owners will have to reapply the pesticide. There are some pest control treatments like lawn granule pellets, that actually work better in the rain.

15. What is the overall concentration of pesticide?

Overall, the concentration of pesticides ranged from 1.0 $\mu g\ kg^{-1}$ to 251 $\mu g\ kg^{-1}$, with a mean of 16 $\mu g\ kg^{-1}$.

16. How do you use Chlorpyriphos 20 EC?

It is effective on Paddy, Beans, Grams, Mung, Sugarcane, Ground Nut, Cotton, Brinjal and other fruits and vegetables. Additionally, CP Gold is also recommended for Termite control at a dosage of 1% on buildings. Dose - 300-500 ml per acre depending on the severity of the pest attack. Soil Treatment - Wheat: 2-3 lit./ha.

17. How long after spraying insecticide is it safe?

What's more, there is no scientific standard for how long one should stay off a lawn after it is treated. Many companies that use these chemicals warn that people should stay away from sprayed surfaces for six to 24 hours.

18. How long does insecticide last?

In general, insecticide manufacturers recommend disposing of their products after two years and usually won't guarantee effectiveness for any longer than two years.

19. How long does insecticide take to dry?

Generally 2 to 4 hours or until dry. Your technician will inform you in advance of any precautions required by the label and our safety policy.

20. What is the full form of WDG?

The full form of WDG is Water Dispersible Granule

21. How much Sevin would be needed per ha, if the dosage is 2 kg a.i. /ha and the local supply store sells 50% WP?

Sevin needed per ha = 4 kg

22. Give two examples of synthetic insecticide?

Parathion, Permethrin

23. Write down the objective for calculation of doses of insecticide?

To know the exact dose required for the preparation of spray fluid to be applied.

24. Name a diluent that is used for special application?

Deodorized kerosene is used as diluent for special application.

25. Why same insecticides are available in market in different concentrations?

Farmers buy insecticides in ready to use form. So, various concentrations of same insecticides are available in the market as per farmer's need.

26. Give some examples of insecticides?

Boric acid, silica gel, diatomaceous earth, malathion, chlorpyriphos, DDT, BHC, permethrin, carbofuran, nicotine, pyrethrum etc.

27. What do you mean if the dosage of insecticides is too low?

Lack of efficient control, waste of money, promotion of resistance ere shown if the doses are too low.

28. What is the formula are used only to calculate when the formulations like EC, SL or WP?

$N_1V_1=N_2V_2$

N_1= Concentration of commercial formulation in percent or grams

V_1= volume or amount of commercial formulation required in milliliter or grams

N_2= Desired or amount of spray fluid in percent

V_2= Volume or amount of spray fluid required (In ml)

29. What is the Objective for the calculation of doses of insecticides?

To know the exact dose required for the preparation of spray fluid to be applied.

30. What is the percentage of purity in formulation of insecticides?

< 80 %

31. What is formulation of insecticides?

A pesticide formulation typically consists of an active ingredient, plus several inactive materials called adjuvants, or additives. Pesticides are available in various "formulations". Some insecticide formulations include dusts, gels, granules, liquids, aerosols, wettable powders, concentrates, and pre-mixed solutions.

32. Why right doses of insecticides are needed to be applied?

Only the correct dosage guarantees optimum pest control whilst also keeping all risks to a minimum.

MCQs

1. Pesticide sprays are available as

i) Wettable powders   ii) Soluble powders

iii) Liquid concentrates   iv) All of the above

Ans: iv) All of the above

2. Which of these are applied without dilution?

i) Dusts   ii) Dusts and granules

iii) Lime   iv) Granules and lime

Ans: ii) Dusts and granules

3. Amount of active ingredient in liq. concentrate is expressed in

i) Percent by weight   ii) Pounds per gallon

iii) Kilogram per hectare   iv) None of the above

Ans: ii) Pounds per gallon

4. Scales are often more practical for

a) Volumetric measures   b) Area measures   c) Mass measures

Ans: a) Volumetric measures

5. Large volume recommendations can be converted to smaller quantities by coupling with

i) Complex algorithm   ii) Data analysis

iii) Statistics   iv) Simple arithmetic

Ans: iv) Simple arithmetic

6. Commercial Sevin WP contains 50 % a.i. How much Sevin is required to make 200 L of 0.09% spray solution?

a) 201 g   b) 205 g   c) 211 g   d)360 g

Ans: d

7. Dimecron 50 EC is recommended to control stem borers at the rate of 1.75 kg a.i. ha^{-1}. How much Dimecron will be required to treat 1250 m^2?

a) 430 mL.   b) 431 mL.   c) 350 mL.   d) 437 mL.

Ans: d

8. How much Furadan (8 % a.i.) is required to spray 8000 m² at 2 kg a.i. ha⁻¹?

a) 30 kg.   b) 40 kg.   c) 20 kg.   d) 50 kg.

Ans: c

9. Rogor 30 % EC is to be sprayed at the rate of 1 L a.i. ha⁻¹ to control thrips. How much Rogor is required to spray 2500 m² ?

a) 0.85 L.   b) 0.83 L.   c) 0.97 L.   d) 1.0 L

Ans: b

10. Monocrotophos (Nuvacron) is recommended to control leaf miner on groundnut at 0.75 kg a.i. ha⁻¹. The commercial product has 36 % EC. How much chemical is required to spray 2 ha?

a) 4.8 L.   b) 5.0 L.   c) 3.2 L.   d) 4.5 L.

Ans: b

11. To control *Helicoverpa* in pigeonpea 400 L of 0.3% solution is to be prepared with Quinalphos (25%a.i. in EC). How much chemical is required to prepare the spray solution?

a) 4.8 L.   b) 5.0 L.   c) 3.2 L.   d) 4.5 L.

Ans: a

12. To control head bugs in sorghum, 300 L of 0.33% carbaryl 50 WP is to be prepared. How much chemical is required for the spray solution?

a) 1.98 kg.   b) 2.0 kg.   c) 1.5 kg.   d) 2.1 kg.

Ans: a

13. To control shoot fly on sorghum 0.35% spray solution is to be prepared with Endosulfan 35 EC. How much chemical will be required for 250 L spray solution?

a) 2.7 L.   b) 2.5 L.   c) 3.1 L.   d) 5.1 L.

Ans: b

14. The leafminer is a pest of

a) millet.   b) castor.   c) groundnut.   d) pigeonpea.

Ans: c

15. The pest that transmits the bud necrosis virus in groundnut crop is

a) shootbug   b) leafminer.   c) spodoptera.   d) thrips.

Ans: d

16. The pest that causes dead hearts in sorghum at the seedling stage is/are

a) aphids.   b) shoot fly.   c) mites.   d) midge.

Ans: b

17. While calculating insecticide doses, we have to consider –

a. Quantity of water   b. Area to be treated   c. Both of these   d. None of these

Ans: c. Both of these

Fill in the blanks

1. Diluents such as ……………… may be used for general applications.

Ans: deoderized kerosene

2. The amount of …………… in liquid concentrates is expressed in pounds per gallon.

Ans: active ingredient

3. If another diluent is used instead of water, the weight per gallon of that diluent should be substituted for ……………………

Ans: 8.3

4. ……………… formulation is generally used for malaria control and residual surface treatment.

Ans: Wettable Powder

5. ……………… effect is seen in Ultra Low Volume Liquid.

Ans: Knockdown

6. ……………… formulation contains the active ingredient (2.5 to 25%) mixed with a petroleum-based solvent and an emulsifier.

Ans: Emulsifiable concentrate

7. Percentage of purity in insecticide formulation is ………………

Ans: < 80 %

8. Active ingredient is mixed with an inert carrier powder like talc or gypsum in …………….. Powder.

Ans: Dustable

9. Organic insecticides are classified into …………….. groups.
Ans: 4

10. Full form of EC is ………………………………..
Ans: Emulsifiable Concentration .

11. Pyrethrum is a plant based ……………….. insecticide.
Ans: Natural

12. Wettable powders contain ………………. active ingredients .
Ans: 10 % -50 %

13. Pesticides for use in spray are generally used as – …… powder and ………. powder.
Ans: wettable, soluble

14. Liquid concentrates of insecticides formulations are used as ……………..
Ans: Spray

15. Before application, insecticides are diluted with ……………..
Ans: water

16. The main purpose of pesticide application technique is to cover the target with ………………….. efficiency and minimum efforts to keep the pest under control as well as minimum contamination of ……………………..
Ans: maximum, non-targets

True-False

1. Frequently, pesticide recommendations are given only for small area applications.
Ans: False

2. The precise amount of water applied to an acre or any other given area is not at all a matter of concern.
Ans: False

3. We can determine the dosage or rate of active ingredient if we know the area covered by a given amount of spray.

Ans: True

4) Forecasting of a pest incidence depends on the catch of insects in the traps (pheromone and light traps).

Ans: True

5. Decision making on a spray operation depends on the degree of infestation and threshold levels of pest and disease incidence.

Ans: True

6. When the walking speed of the person operating the sprayer increases, the spray liquid ha^{-1} decreases.

Ans: True

7. Acaricides are useful to kill mites

Ans: True

8. Fumigants are chemicals that enter the body of the insect by body contact.

Ans: False

9. Emulsifiable concentrate is very sensitive to cold during storage.

Ans: True

10. Wettable powder is expensive.

Ans: False

11. Dust formulation contains 1% to 15 % active ingredient .

Ans: False

12. Formulation = Pure form (active ingredient) + additives.

Ans: True

13. If insecticide dosage is low, it means promotion of resistance.

Ans: True

14. Malathion is a Carbamate compound.

Ans: False

15. Amount of active ingredient determines the amount of insecticide to be applied.

Ans: True

16. Active ingredient is present in insecticidal formulation in very least amount.

Ans: True

17. Calculation of doses of pesticides, insecticides is not necessary.

Ans: False

18. For organizations with a number of stores, it is advisable to introduce standardized stack sizes and issue technical instructions for treatments.

Ans: True

Chapter - 11

Plant Protection Equipments and Spray Droplet Size

Introduction

Chemicals are widely used for controlling diseases, insects and weeds in the crops. They are able to save a crop from pest attack only when applied in time. They need to be applied on plants and soil in the form of spray, dust or mist. The chemicals are costly. Therefore, equipment for uniform and effective application is essential. Dusters and sprayers are generally used for applying chemicals. Dusting, the simpler method of applying chemical, is best suited to portable machinery and it usually requires simple equipment. But it is less efficient than spraying, because of the low retention of the dust. High volume spraying is usually effective and reliable but is expensive. Low volume spraying to some extent overcomes the failings of each of the above two methods while retaining the good points of both.

Spraying is employed for a variety of purposes such as application of:

i.   Herbicides in order to reduce competition from weeds.

ii.   Protective fungicides to minimize the effects of fungal diseases,

iii. Insecticides to control various kinds of insects' pests.

iv.   Micro-nutrients such as manganese or boron.

The main function of a sprayer is to break the liquid into droplets of effective size and distribute them uniformly over the surface or space to be protected. Another function is to regulate the amount of insecticide to avoid excessive application that might prove harmful or wasteful. A sprayer that delivers droplets large enough to wet the surface readily should be used for proper application. Extremely fine droplets of less than 100-micron size tend to be diverted by air currents and get wasted. Crops should, as far as possible, be treated in regular swaths. By use of a boom, uniform application can be obtained with constant output of the machine and uniform forward travel.

Types of sprayers

Different designs of spraying equipment have been developed for different types of applications, field and crop conditions. Manually operated hydraulic sprayers viz. Knapsack sprayers, twin knapsack sprayers, foot sprayers, hand compression sprayers; air carrier sprayers such as motorized knapsack mist blower cum duster (LV) and centrifugal rotary disc type sprayers are specially suitable for spray applications in crops. The present trend is to apply concentrated pesticides by means of low and ultra low volume sprayers. This has been possible through the development of better formulations and special nozzles. New Controlled Droplet Application (CDA) atomizers require less than 15 1/ha of spray mixture and are easy to operate. Although new type of spray atomizers is available but the correct chemical formulations are commonly not produced. Hence their use is limited to particular crop, pest or disease. The instructions of the manufacturer should be carefully read whether a formulation is recommended for ultra low or low volume application.

1. Manually operated sprayers

a. Hydraulic sprayers

i. Hand sprayer (Kitchen sprayer): The tank capacity is 0.5 – 1 litre. It is commonly used for spraying in kitchen gardens.

ii. Hand syringe: It consists of a cylinder. It is useful to operate in a small area.

iii. Foot sprayer: It is used to spray tall crops upto 4 meters.

iv. Knapsack sprayer: It is used for spraying low crops, vegetables, nursery, and field crops. The capacity of this sprayer is generally 15 litres.

v. Bucket pump sprayer

vi. Rocker sprayer

vii. Wheel barrow sprayer

b. Compression sprayers

i. Pneumatic hand sprayers

ii. Pneumatic knapsack sprayer

c. Air blast sprayers (Hand atomizer)

2. Power operated sprayer

a. Hydraulic power sprayer

b. Spray blowers

c. Electrodyne sprayer

d. Controlled droplet application (CDA) sprayer

Types of dusters

Manually operated dusters (Hand operated)

a. Rotary type

b. Plunger type

c. Bellows type

d. Traction type

A. Power operated dusters

Hand Sprayer

The hand sprayer is a small capacity pneumatic sprayer. It consists of chromium plated brass tank having a capacity of 0.5 to 3 litres (one litre is more common) which is pressurized by a plunger pump. The air pump remains inside the tank. The sprayer has a short delivery tube to which a cone nozzle is attached. In some models, the nozzle is attached at the top of the tank with flow spring actuated lever, which regulates the flow of the spray liquid. For spraying, the tank is usually filled to three-fourths capacity and pressurized by air pump. The compressed air causes the agitation of the spray liquid and forces it out, on operation of the trigger or shut off type valve. Usually, the chemicals with suspension characteristics cannot be effectively sprayed with this type of sprayer. For spraying wettable powders, the sprayer is shaken frequently to prevent settling of the chemical. For operation, the spray nozzle is directed to the target after charging. It is fitted with mist spray nozzle with gooseneck bend. The pump assembly is made of brass and operated by one person.

Specification

Diameter of the tank	130 mm
Height	210 mm
Weight	1.2 kg

Uses

It is ideal for small nurseries, rose plants, kitchen gardens and spraying wettable insecticides and fungicides.

Stirrup Pump Sprayer

The stirrup pump sprayer commonly used for mosquito control, is quite popular among the small farmers and vegetable growers due to its simplicity, low cost and ease of operation. It is also called bucket sprayer, as the pump always remains submerged in a bucket containing the spray liquid. It consists of a double action pump, seamless brass tube barrel, adjustable stirrup foot rest, angular delivery spout, relief valve in the foot valve, plunger, shaft fitted with travel limiter, D type handle, brass balls, detachable goose neck bend in the spray lance, spray lance fitted with nozzle, delivery hose and trigger cut- off valve with strainer. For operation, the barrel fitted with pump is placed in a bucket containing spray liquid and one person operate the pump by placing his foot on the foot stirrup and moves the pump shaft fitted with D type handle. The other person holds the spray lance and directs it towards the target. All the major parts are made of brass. Usually a flat fan spray nozzle is used with the sprayer.

Specification

Delivery pipe	5 m
Field capacity of the sprayer	0.3 ha/day

Uses

It is used for spraying in orchards, nurseries, flower crops, vegetable gardens etc.

Hand Compression Sprayer

Hand compression sprayers are either pressure retaining or non-pressure retaining type. The pressure retaining type has an advantage that air charged once may last for weeks, but requires sturdy tank and high pressure, therefore these are not in common use. Non-pressure retaining type is the most commonly used hand compression sprayer. Like other sprayers, it consists of an airtight metallic tank, air pump, lance fitted with trigger type or shut off valve, gooseneck bend a pair of shoulder mounted straps and nozzle. It has to carry it on the back. All the parts are made from brass alloy and the tank is fabricated to withstand high-pressure up to the order of 18 kg/cm². For operation, the tank is filled to three fourths of its capacity and pressurized by hand plunger pump, which remain inside the tank or from a compressor. The pressure inside the tank is usually maintained at 3-4 kg/cm². The operator mounts the sprayer on his back securing it by shoulder straps and operates the trigger valve, which enables the spray liquid to flow through lance and nozzle. The lance is directed towards the target. A single person can operate the sprayer. For maintaining proper atomization of the spray liquid, the tank requires frequent pressurization. The discharge and atomization decreases with decrease in pressure.

Specifications

Diameter of the tank	200. mm.
Height of the tank	670 mm
Capacity of the tank	6, 8, 12, 14, 16 (in litre)
Weight	7.5 kg
Field capacity	0.4 ha/day

Uses

The hand compression sprayer is used in kitchen gardens, nurseries, vegetable gardens, flower crops and field crops.

Rocker Sprayer

The rocker sprayer is a long lever high-pressure sprayer designed for operation with one or two lances. The complete assembly is mounted on a wooden board, which is held to the ground by the foot of the operator. The sprayer consists of a single or double acting piston pump for developing high pressure, an air chamber, spray lance with shut off valve and strainer, 5 m suction line fitted with strainer and delivery line. The principal components are made from brass alloy. The lance is fitted with gooseneck bend and nozzle and the length of lance may vary from 60 to 90 cm. The pump is operated with long lever to and fro in a rocking motion which suck the liquid from the inlet pipe submerged in the spray liquid. The other person holds the lance and directs the spray chemical to the target. If two lances are used, then it may require in all three persons for the spraying operation. With high jet spray gun or bamboo lance the spray chemical can be delivered to a height of up to 10 m.

Specifications

Overall length	760 mm
Overall width	150 mm
Overall height	1230 mm
Weight	6 kg
Pressure developed	10 kg/cm
Field capacity	1 ha/day (with 1 lance) and 1.5-2 ha/day with 2 lances

Uses

For spraying on tall trees like coconut, arecanut, sugarcane, rubber plantations, orchards, vineyards and field crops, vegetable gardens, flower crops etc.

Foot Sprayer

The foot sprayer is one of the ideal and versatile sprayers used for multipurpose spraying jobs. The principle of working is similar to the rocker sprayer. The sprayer consists of a pump operated by the foot lever, suction hose with strainer, delivery hose, spray lance fitted with shut off pistol valve, gooseneck bend and adjustable nozzles. The pump barrel is mounted on steel frames, which provide it stability when placed on the ground. It has a provision of two strong springs, which retract the foot lever to its original position after each pumping stroke. The sprayer does not have inbuilt tank, therefore an additional storage device or container is required to store the spray liquid in which the strainer of suction hose remain submerged. It has provision for the two discharge lines, which increases its versatility and field capacity. The plunger pump being a positive displacement pump, builds up a high pressure to throw spray liquid to larger distances with a suitable boom. The pump barrel, lance and the spray nozzle are made from brass alloy. For operation the inlet pipe is placed in the storage container and one person continuously operates the pump by foot lever. There is a provision for the operator to hold the sprayer at the top by U-type fixture. The other person directs the lance to the target. For spraying tall trees up to a height of m, a high jet or bamboo lance can be used.

Specifications

Overall length	440 mm
Overall width	170 mm
Overall height	980 mm
Weight	10.0 kg
Pressure developed	14-18 kg/cm^2
Field capacity	1 ha/day (with 1 lance) and 1.5-2.0 ha/day (with 2 lances)

Uses

The foot sprayer is all purpose sprayer, suitable for both small and large-scale spraying on field crops, in orchards, vegetable gardens, tea and coffee plantations, rubber estates, flower crops, nurseries etc.

Knapsack Sprayer

Knapsack sprayer consists of a pump and a air chamber permanently installed in a 9 to 22.5 liters tank. The handle of the pump extending over the shoulder or under the arm of operator makes it possible to pump with one hand and spray with the other. Uniform pressure can be maintained by keeping the pump in continuous operation.

Specifications

Person required	One person
Tank capacity	9 - 22.5 litre
Pump cylinder inner diameter	39-42 mm
Number of piston in pump cylinder	One
Pressure chamber capacity	572-660 ml
Displacement volume	87.24 ml
Type of delivery spout	Threaded
Cut off valve passage diameter	5 mm
Lance length	725 mm
Nozzle type	Hollow cone
Spray angle	78 degree
Size of filling hole	94.9 mm
Pump discharge	610-896 ml/minute

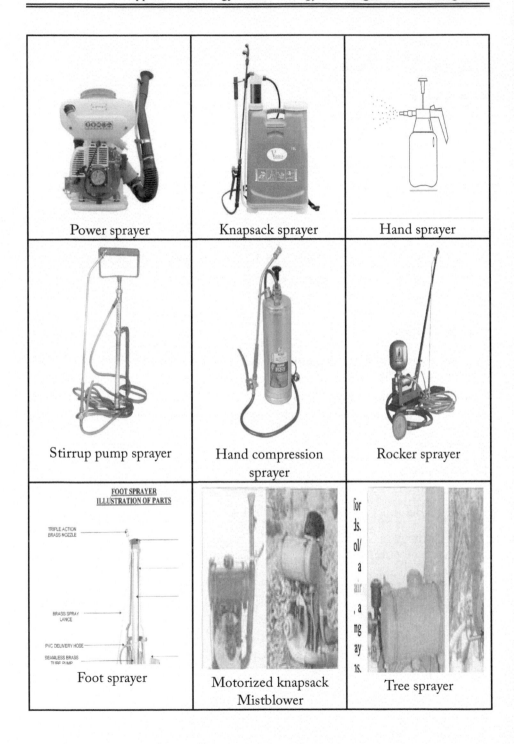

Power sprayer	Knapsack sprayer	Hand sprayer
Stirrup pump sprayer	Hand compression sprayer	Rocker sprayer
Foot sprayer	Motorized knapsack Mistblower	Tree sprayer

Uses

Knapsack sprayers are used for spraying insecticides and pesticides on small trees, shrubs and row crops.

Power Sprayers

Power sprayers are used for developing high pressure and high discharge for covering large area. These sprayers are either operated by auxiliary engines or electric motors. Most of these sprayers are hydraulic sprayers and consist of power unit to drive the pump, pump unit which employs piston or plunger pump, piston (1 to 3), pressure gauges, pressure regulators, air chamber, suction pipe with strainer, delivery pipes fitted with lance, gooseneck bend and nozzles. The portable sprayers use petrol engine so that these can be easily taken to the spray sites. The complete assembly is mounted on the stretcher type frame or on wheel barrow for easy transportation. The number of lances may vary from I to 6 depending upon the model. In some models there is a built in storage tank of fibre glass having capacity of 100 litres, while in others a separate storage tank is required in which the suction pipe of the sprayer remains submerged. For operation, the shut off trigger valve of the lance is closed and the engine/ electric motor is started to actuate the pump. The pump draws the spray liquid from the tank, imparts pressure energy and sends it to the delivery line/lines. The operator directs the lance towards the target and operates the trigger/shut off valve. Adjusting the nozzle or selecting the appropriate nozzle, adjusts the spray pattern. For delivering the spray liquid to large distances/ height a bamboo lance can also be used.

Specifications

Overall length	740 mm
Overall width	610 mm
Overall height	990 mm
Number of spray lance	upto 6
Discharge	up to 2.5 litre/min
Field capacity	0.2-0.3 ha/h
Weight	6.60 kg
Pressure developed	up to 40 kg/cm^2
Power required	3 hp

Uses

These sprayers are suitable for spraying in orchards, tea and coffee plantations, rubber plantations, vineyards and field crops. Tall trees up to a height of 15 meters can be sprayed with these types of sprayers.

Motorised Knapsack Mist Blower

The motorized knapsack mist blower has a small 2-stroke petrol/ kerosene engine of 35 cc to which a centrifugal fan is connected. The centrifugal fan is usually mounted vertically. The fan produces a high velocity air stream, which is diverted through a 90-degree elbow to a flexible (plastic) discharge hose, which has a divergent outlet. The spray tank that has also a compartment for fuel and engine-fan unit is mounted on a common frame, which fits to the back of operator. The tank is made of plastic. The spray liquid flows due to gravity and suction created at the tip of nozzle, thus remains in the air stream. Some models of the sprayers have roller pump for pumping the spray liquid into the discharge hose and have tall tree spraying attachment. These sprayers have shear type nozzles. For operation, the tank is filled with the spray liquid and cranking with the rope starts the engine. Upon rotation of the engine the fan produces a high velocity air stream. The control valve for the spray liquid is opened gradually and adjusted for the desired flow rate. The operator directs the discharge hose to the target. The spray liquid that falls in the air stream gets sheared and upon coming in contact with the atmosphere a mist is produced.

Specifications

Overall length	460 mm
Overall width	220 mm
Overall height	670 mm
Weight	1 1.0 kg
Air velocity at the outlet	65-75 m/s
Power	1.2 hp
Fan output	7 m³/min
Field capacity	3 ha/h

Uses

It is used for spraying in orchards, coffee estates and tall crops.

Tree Sprayers

The tree sprayer is an ideal sprayer for spraying tall fruit trees in the orchards. It consists of a 4- stroke petrol/ kerosene engine to drive the fan, a centrifugal fan which produces stream of high volume and velocity, a micronized nozzle for producing uniform and fine droplets of spray liquid in the range of 150-200 microns, plastic tank for storage of spray liquid, rotary pump to draw the spray liquid from the tank and to feed it to the nozzle and a fibre glass casing. All these components are joined and mounted on the stretcher type of frame. The sprayer can be carried by two persons to the place of spraying. For operation, the tank is filled with spray liquid and the engine is started by cranking with the rope, to drive the fan and the rotary pump. The control valve is opened to adjust the rate of flow. The sprayer is placed under the tree and manually moved around it to complete the spraying operation.

Specifications

Power	3 hp
Fan output	1000 m³/h
Tank capacity (l)	10, 90 litre
Vertical range (m)	20-25 m
Field capacity (ha/day)	3-4 ha/day

Uses

The sprayer is used for spraying of tall fruit trees.

Knapsack Power Sprayer

The pump adopts horizontal gear driving. It is powerful and stable in pressure. As it is oil-soaking type, the lubricating effect of crank box is positive. It adopts double cylinder pump, which raises operational efficiency. The piston is heat treated and wear resistant. 'V 'packing adopts special materials so it is durable. The engine has electronic ignition, which is easy to be operated and maintained. Its engine has high power to weight ratio. It uses gasoline as a fuel and produces one power stroke for each 180^0 of crank rotation. Pressure is controlled by oil valve; spraying pressure is changeable. Its structure is rigid; materials are excellent and easy to be maintained.

Specification

Length	650 mm
Width	450 mm
Height	570 mm
Weight	9.8 kg
Number of gangs	2
Plunger	1 No. (Double acting)
Cylinder	2 Nos.
Diameter of plunger	16 mm
Stroke	8 mm
Spraying volume	4.8 — 5.2 litres
Pressure	20-25 kg/cm^2

Uses

It is suitable for spraying pesticides and fungicides on rice, fruits and vegetable crops.

Lower Sprayer

These are tractor mounted or trailed air carrier type of sprayers for spraying in orchards. The spray volume can be controlled making it high volume, low volume or ultra low volume sprayer depending upon the requirement. The sprayer mainly consists of highpressure piston pump, which atomise the spray solution, axial or centrifugal fan to produce a stream of air, interchangeable and orientable spray heads, high capacity spray tank and other control attachments. All the sub-assemblies are mounted on the frame and can be attached to 3-point linkage of the tractor or mounted on a trailer. The spray heads are mounted radially so as to cover the 2-rows of the trees. The spray heads can be oriented upwards for spraying the tall trees. In case of spraying individual tree the spray heads can be exactly adjusted to the form of tree. The volume of spray and the air stream velocity and volume can also be adjusted as per the requirement. For operation the spray liquid is filled in the tank. The tractor power is used to drive the fan and pump. The pump sucks and sends the liquid to the spray head, which atomise the liquid, and at the same time droplets meet the air stream to carry them to the target.

Specifications

Length	1360 mm
Width	970 mm
Height	2050 mm
Length of propeller shaft	600 mm

Uses

For spraying tall trees, plantation crops, vineyards and other field crops.

Power Tiller Mounted Orchard Sprayer

It consists of an HTP (horizontal triplex piston) pump, trailed type main chassis with transport wheels, chemical tank with hydraulic agitation system, cut off device and boom equipped with turbo nozzles. It is fitted with turbo nozzles with operating pressure of 9-18 kg/cm^2. It generates droplets of 100-150 micron sizes. Depending upon the plant size and their row spacing, the orientation of booms can be adjusted. The spray booms are mounted behind the operator.

Uses

For spraying of foliage, in orchard crops like pomegranate, orange, sweet lime and grapes.

Tractor Mounted Sprayer

These are hydraulic energy sprayers. Basically the spray boom can be arranged in two ways; ground spray boom and overhead spray boom. The overhead spray boom is designed for tall field crops and the planting is done in such a way that it leaves an unplanted strip of about 2.5 m width for operation of the tractor. Therefore a planted strip may be 18-20 m wide and after every planted strip a fallow strip has to be left for tractor operation. For ground spray boom the planting has to be done in rows keeping in view track width of the tractor. It is suitable for use when the crop is small. The sprayer essentially consists of a tank which is made of fibre glass or plastic, pump assembly, suction pipe with strainer, pressure gauges, pressure regulators, air chamber, delivery pipe, spray boom fitted with nozzles. The complete sprayer is mounted on 3-point linkages of the tractor. It uses high pressure and high discharge pump as the number of nozzles may be up to 20 depending upon the crop and make of the sprayer.

Specifications

Overall length (mm)	6340 mm
Overall width (mm)	1290 mm
Overall height (mm)	1570 mm
Tank capacity (l)	400 litre
Weight (kg)	150 kg
Field capacity (ha/day)	8 ha/day (with 14 nozzles)

Uses

It is used for spraying in vegetable gardens, flower crops, vineyards and for tall field crops like sugarcane, maize, cotton, sorghum, millets etc.

Self Propelled Light Weight Boom Sprayer

The machine is operated by 5 hp diesel engine and is controlled by the operator from the handle. The machine consists of a light weight power tiller unit and a spraying unit. Spray pump and two narrow pneumatic wheels get power from the engine through gears, chains and sprockets. One caster wheel is also provided at the rear, which acts as supporting wheel. The spray boom is mounted on the power unit through a canopy frame.

Specifications

Power required	4.8 hp, diesel engine
Engine speed	2600 rpm
Type of pump	Roller type
Ground clearance	500 mm
Boom height	600-1300 mm
Number of nozzles	12
Type of nozzles	Flat fan/any other
Nozzle spacing	500 mm
Tank capacity	100 litre
Pump speed	1600 rpm
Swath	6300 mm

Wheel diameter	440 mm
Tread width	920 mm
Weight	208 kg

Uses

The self-propelled light weight boom sprayer is used for chemical application on wheat, vegetable and other crops

Front Mounted Self-Propelled Boom Sprayer

It consists of chassis with ground drive system, diesel engine (5.5 hp), spray boom with hollow cone nozzles, power transmission system, gear box, lugged cage wheels, chemical spray tank and pump. The pump and boom with 14 nozzles are mounted on a stand, which can be attached in front of the frame.

Specifications

Power source	5.5 hp, Diesel engine
Boom length	6210 mm
Number of nozzles	14
Type of nozzle	Hollow cone
Nozzle pressure	3 kg/cm^2
Distance between nozzles	300-600 mm (adjustable)
Chemical tank capacity	50 litre
Weight (kg)	140 kg

Uses

For uniform spraying in horticultural crops and other crops like cotton, maize, groundnut.

Self Propelled High Clearance Sprayer

The machine has a chassis with 1200-mlli ground clearance, four wheels, and 20 hp diesel engine, gearbox, water tank, seat for the operator, spray pump and boom with 18 nozzles. The nozzles are spaced at 675 mm and total boom width is 10.80 m. The boom height can be adjusted form 31.5 cm to 168.5 cm to suit different crops and can be folded during transport. The front two wheels are narrow in width

(20 cm) and are given drive, while the rear wheels are steering wheels. Fenders have been provided in front of the drive wheels to deflect the crop branches away from the wheels for reducing mechanical damage. The wheel track is 135 cm and during operation 2 rows of cotton crop come under the machine chassis. Machine has four forward speeds and one reverse speed. The 1st and 2nd gears are for field operation while 3rd and 4th gears are for road transport. The field speed is up to 5 km/h and the road speed is up to 25 km/h.

Specifications

Length	4250 mm
Width .	1950 mm
Height	2650 mm
Crops for which machine is suitable	All row crops except paddy
Number of gears	Four forward and one reverse
Ground clearance of machine	1200 mm
Length of boom	10.80 m
Number of nozzles	18
Nozzles spacing	67.5 cm (fixed)
Width of coverage	13.50 m
Tank capacity (litres)	1000 litres
Maximum field speed	5 km/h
Maximum road speed	25 km/h
Weight	1900 kg
Power required	20 hp, diesel engine

Uses

Self-propelled high clearance sprayer is most suitable for spraying on tall crop like cotton. The machine can also be used for spraying on sunflower, wheat and other crops.

Aeroblast Sprayer

The machine consists of tank of 400 litres capacity, pump, fan, control valve, filling unit, spout adjustable handle and spraying nozzles to release the pesticide solution in to stream of air blast produced by the centrifugal blower. The air blast distributes chemical in the form of very fine particles throughout its swath, which is on one side of tractor. The major portion of swath is taken care of by the main blast through the main spout and the auxiliary nozzles cover the swath area near the tractor. The sprayer is mounted on the tractor 3-point linkage. The orientation of air outlet can be adjusted for its direction and width of coverage.

Specifications

Power required	35 hp or above
Crops suitability	Cotton, sugarcane, sunflower and horticulture
Type of blower	Centrifugal
Type of pump	Piston
Flow rate (l/min)	120 litre/minute
Spray rate (l/ha)	100-400 litre/ha
Spray swath (m)	15 m (without wind)
Fan speed (rpm)	3660 rpm
Working speed	2-6 km/h
Tank capacity	400 litre
Machine weight	230 kg

Uses

It is useful for spraying on horticultural trees and crops like cotton, sunflower etc.

Battery Powered Low Volume Knapsack Spinning Disc Sprayer

The sprayer consists of plastic tank, grooved spinning disc, a fractional horsepower DC motor to spin the disc at very high speed, 12 volt power supply source which can be a dry battery or lead acid battery and a light weight handle on which the spray head is mounted. The handle has provision to adjust the angle of the spray head. The spinning disc is also made of fine quality plastic, which has very fine radial grooves. For operation, the spray liquid in minimum dilution is filled in the tank, which flows in the form of drops at the centre of spinning disc. The disc is attached to the motor, which rotates it at a very high speed. The spray liquid travels in the grooves of the disc and gets fragmented into very fine droplets when it leaves the periphery of the disc. The droplets are thrown outwards due to centrifugal force.

Specifications

Dimensions	1810 mm x 660 mm x 370 mm
Weight (kg)	7
Tank capacity (l)	10
Number of spinning discs	One
Power source	6 V rechargeable battery
Application rate (l/ha)	45
Field capacity (ha/h)	0.20
Labour requirement (man-h/ha)	5

Uses

It is suitable for spraying in crops like paddy, cotton, groundnut, pulses and vegetable. It saves 30 per cent labour and operating time and 15 per cent on cost of operation compared to manual spraying. It also results in 27 per cent increase in yield compared to spraying by manual sprayer

Ultra Low Volume (ULV) Sprayer

These are also known as controlled droplet application (CDA) sprayers and can apply correct size of droplets. Due to very low volume application features the spray chemicals can be applied in very low dilution or no dilution at all with these sprayers. Common type of ULV sprayers are hand held and number of these units can be mounted on the boom attached to the tractor to increase the field capacity. The sprayer unit consists of a plastic container having 1-2 litres capacity, grooved spinning disc, a fractional horsepower DC motor to spin the disc at very high speed, 12-volt power supply source which can be air dry battery or a lead acid battery and a light weight handle on which the spray head is mounted. The handle has provision to adjust the angle of spray head. The spinning disc is also made of fine quality plastic, which has very fine radial grooves. For operation, the spray liquid in minimum dilution form is filled in the container, which flow in the form of drops at the centre of spinning disc. The disc is attached to the motor, which rotates at a very high speed. The spray liquid travels in the grooves of the disc and gets fragmented into very fine droplets when it leaves the periphery of the disc. The droplets are thrown due to the centrifugal force. One charge of battery may last up to 15 hours of spraying. Fan assisted ULV sprayers are also available which can be effectively used in green houses or glass houses.

Specifications

Overall length	1950 mm
Overall width	250 mm
Overall height	280 mm
Weight	8.5 kg
Droplet size	35-100 micron
Application rate	1-5 l/ha
Field capacity	2-3 ha/day having spray width of 1 m

Uses

It can be used for spraying in nurseries, vegetable crops, vineyards and other field crops.

Hand Rotary Duster

The hand rotary duster is available in two models, shoulder mounted and belly mounted. It is a common type of duster being used by the farmers. The duster consists of a hopper, fan/blower, rigid/flexible discharge pipe, reduction gearbox, rotating handle, shoulder straps, and metering mechanism. The hopper is either made of plastic or aluminium. The hopper made from mild steel sheet is coated with anti-corrosive material for longer life. The duster has mechanical agitator connected to the gearbox placed in the hopper, which churns the chemical and prevent clogging of the outlet. The adjustable orifice plate mounted below the hopper outlet controls the application rate. The fan/blower is enclosed in the casing and is rotated with the handle through gearbox. For operation, the hopper is filled 1/2 to 3/4th of the capacity of the hopper. This is mounted on the shoulder/belly with the help of adjustable straps. The discharge pipe fitted with spoon type deflector is directed towards the target continuously rotating the handle. The chemical in dust/powder form drops from the hopper in the discharge pipe having an air stream created by the blower. These dust particles emerging in the form of cloud from the discharge pipe are carried to the plant where these settle on the leaves, stems and other parts.

Specifications

Overall length	280 mm
Overall width	330 mm
Overall height	330 mm

Weight	3.5 kg
Hopper capacity	5 litre
Field capacity (ha/ day)	0.6 ha/day (for field crops)

Uses

For control of pests and diseases by use of chemicals in the dust forms in nursery, vegetable gardens, field crops, tea and coffee plantations, green houses, glasshouses and go downs.

Types of nozzles

A. Hydraulic nozzle

    a.   Flood jet
    b.   Flat fan
    c.   Cone

        i.   Hollow cone
        ii.   Full cone

B. Gaseous nozzle

C. Centrifugal nozzle

D. Kinetic/Vibrajet nozzle

E. Thermal or hot tube nozzle

Parts of nozzle

a. Body
b. Cap
c. Swirl plate
d. Orifice plate
e. Washer
f. Stainer

Parts of a Knapsack sprayer

a. Container or tank
b. Spray pump
c. Agitator
d. Filter

e.   Hose pipe
f.   Spray lance
g.   Cut off valve
h.   Boom
i.   Nozzle

Pesticide application methods:

It depends on nature of pesticide, type of formulation, type of pests, site of application, availability of water etc.

a.   Dusting

b.   Spraying: Spraying techniques are classified as high volume (HV), low volume (LV) and ultra low volume (ULV), according to the total volume of liquid applied per unit of ground area. Initially high-volume spraying technique was used for pesticide application but with the advent of new pesticides the trend is to use least amount of carrier or diluents.

c.   Granular application: i) Broadcasting ii) Infurrow application iii) Side dressing iv) Spot application v) Ring application vi) Root zone application vii) Leaf whorl application viii) Pralinage

d.   Seed dressing or pelleting

e.   Seedling root dip

f.   Sett treatment

g.   Trunk or stem injection

h.   Swabbing

i.   Root feeding

j.   Soil drenching

k.   Baiting

l.   Fumigation

Planning of pesticide application

It should be done well in advance with the following considerations

   a. Formulation of the application programme

b. Training of the operators

c. Checking of equipments

d. Stocking of pesticides

e. Arrangement of water

f. Compatibility of pesticides

g. Consideration of weather

h. Pre sowing/pre emergence/post emergence application

i. Record keeping

Calibration of pesticides application equipments

Calibration is most important to secure correct dose application in an identified area in the available time as per recommendations to achieve effectiveness. The purpose of calibration is to ensure that the equipment delivers the correct amount of pesticide uniformly over the target area. Too much application and too less application of water are not desirable for effective pest management. Too much application of water per unit time as well as per unit area leads to wastage of pesticide, crop injury and uneconomical. Similarly if less water is applied in the same condition, there will be problems like wastage of pesticide, poor pest control, wastage of time and money etc.

Volume of spray mixed applied per unit area depends on

a.　Nozzle spray discharge rate

b.　Swath width

c.　Walking speed of operator

Field sprayer calibration: It includes

a.　Checking of driving speed

b.　Selection of perfect nozzle

c.　Checking of liquid system

d.　Checking of nozzle output

Measuring of driving speed

a.　Half fill the spray tank with water

b.　Mark out 100 m – note time to drive the distance

$$\text{Driving speed (km/h)} = \frac{\text{Distance driven (m)} \times 3.6}{\text{Time (sec)}}$$

Eg: If it takes 50 seconds to drive 100 meters then the spraying speed is 7.2 km/hour.

Measuring nozzle output

> » Set the pressure

> » Adjust the pressure equalizing valves

> » Measure the nozzle output for one minute

> » Repeat this- measure at least 2 nozzles for every boom section

> » Calculate the average nozzle output

Mark an area of 2.5 m x 4 m. Fill the container and the discharge line of the sprayer with water. Mark the level of the water in the container and spray the area. Now refill the container to the original level with a known volume of water. Alternatively, the volume of water used to spray the above area can also be calculated from the time taken to spray the area and the discharge of the nozzle per unit of time. In each case, the rate of spray delivery in l/ha can be determined by multiplying the volume of water in liters used in spraying by 1000.

$$F = \frac{SDA}{10000}$$

Where,

F — Flow rate in L/min

S — Swath width in meter

D — Operator's walking speed in m/min

A — Application rate in L/ha

Problem: A knapsack sprayer discharges 600 ml liquid every minute and sprays one meter swath. If the operator walking speed is 30 m/min, what is the rate of application in L/ha?

$$F = 600 \ ml \ / \ min \ = 0.6 \ L \ / \ min$$

$$S = 1 \ meter$$

$$D = 30 \ m/min$$

$$A = ? \ L/ha$$

$$F = \frac{SDA}{10000}$$

$$\Rightarrow A = \frac{F \times 10000}{SD}$$

$$\Rightarrow A = \frac{0.6 \times 10000}{1 \times 30}$$

$$\Rightarrow A = 200 \ L \ / \ ha$$

Spray Droplet Size

In spraying, the optimum droplet size differs for different types of application. Fine droplets are required to control insects, pests or diseases and bigger size droplets for application of herbicides, etc. The greater the number of fine droplets produced by the device better will be deposition on target area. The size of droplet is important as it affects drift and penetration distance of droplets towards the target. Hence a compromise is to be made to prevent drift, achieve wide coverage of plant or target area and more penetration. The optimum droplet sizes are indicated in below:

Optimum droplet sizes for different targets

Target group	Droplet size (microns)
Flying insects (drift)	10-15
Crawling and sucking insect (drift)	30-50
Plant surfaces (limited drift)	60-150
Soil application (no drift) as in case of herbicide application	250-500

MCQs

1. Plant protection equipments are used in which form?

a) Spray   b) Dust   c) Mist   d) All the above

Answer: d) All the above

2. Sprayer can be used to apply_____.

a) Herbicide   b) Insecticide   c) Fungicide   d) All the above

Answer: d) All the above

3. How much spray solution (in litre) to be used per hectare in case of field crops and orchards respectively in case of high-volume sprays?

a) 1500-2000 and 500-1000   b) 500-1000 and 1500-2000
c) 500-1500 and 1000-2000   d) 1000-2000 and 500-1500

Answer: b) 500-1000 and 1500-2000

4. Which volume sprays are known as concentration sprays?

a) High volume sprays   b) Ultra-low volume sprays
c) Low volume sprays   d) None of the above

Answer: c) Low volume sprays

5. In ultra-low volume sprays, size of droplets varies between _____.

a) 10-100 micron   b) 20-100 micron
c) 15-100 micron   d) 40-60 micron

Answer: b) 20-100 micron

6. In battery operated sprayers, the component which breaks the chemical solution into fine particles is called _____.

a) Spray gun   b) Nozzle disc
c) Cut-off valve   d) None of the above

Answer: b) Nozzle disc

7. Which of the following component of the sprayer is most important?

a) Cut-off lever   b) Spray gun
c) Nozzle   d) Strainer

Answer: c) Nozzle

8. Which type of nozzle is used to cover a large area at a small range?

a) Hollow cone nozzle   b) Solid cone nozzle
c) Fan nozzle   d) All the above

Answer: b) Solid cone nozzle

9. Which type of nozzle is used for low pressure spraying?

a) Hollow cone nozzle   b) Solid cone nozzle
c) Fan nozzle   d) None of the above

Answer: c) Fan nozzle

10. Mark the odd one out with respect to absence of built-in pump:

a) Knapsack sprayer   b) Hand atomizer
c) Hand compression sprayer   d) Rocker sprayer

Answer: d) Rocker sprayer

11. Capacity of the tank of knapsack sprayer is _____ litres.

a) 5- 10   b) 10-20   c) 15-20   d) 20-40

Answer: b) 10-20

12. Which sprayer has coverage rate 0.5-1 ha/day?

a) Rocker sprayer   b) Foot or pedal sprayer
c) Knapsack sprayer   d) Hand atomiser

Answer: c) Knapsack sprayer

13. How much power is required to generate the engines of motorised knapsack sprayer and power sprayer respectively?

a) 1.2-3 hp and 3 hp   b) 1.2-2 hp and 3hp
c) 3 hp and 1.2-3 hp   d) 3hp and 1.2-2 hp

Answer: a) 1.2-3 hp and 3 hp

14. Essential component of duster is _____.

 a) Agitator   b) Hopper
c) Delivery tubes   d) Rotary fan

Answer: b) Hopper

15. Which of the following duster is used in kitchen garden?

a) Rotary hand duster   b) Bellow duster
c) Plunger duster   d) All of the above

Answer: c) Plunger duster

16. Hand rotary duster is also known as _____.

a) Crank duster   b) Fan type duster
c) Both of them   d) None of them

Answer: c) Both of them

17. Which is the most commonly used duster in India?

a) Plunger duster   b) Knapsack duster
c) Rotary hand duster   d) Bellow duster

Answer: b) Knapsack duster

18. Which duster allows prompt coverage of large areas efficiently?

a) Aerial duster   b) Knapsack duster
c) Bellow duster   d) Rotary duster

Answer: a) Aerial duster

19. In which volume spray, water is the carrier of toxicant?

a) High volume spray   b) Low volume spray
c) Ultra-low volume spray   d) More than one option

Answer: d) More than one option

20. What is the droplet size (in micron) for insects on foliage?

a) 10-50   b) 30-50
c) 40-100   d) 250-500

Answer: b) 30-50

21. Plant protection equipments are used to apply

(a) insecticide   (b) herbicide
(c)none of the above,   (d) both of the above.

Ans: (d) both of the above

22. Hand sprayer is filled up to:

(i) ¾ th capacity   (ii) 2/4 th capacity,

(iii) ¼ capacity   (iv) 100%capacity

Ans: (i) ¾ th capacity

23. Which of the following is a part of fogging machine:

(i) fogging coil   (ii) pump

(iii) fogging nozzle   (iv) all of the above

Ans: (iv) all of the above

24. No of crank rotation produced in one power stroke of knapsack power sprayer:

(i) 2000   (ii) 1800   (iii) 1500   (iv) 1300

Ans: (ii) 1800

25. The length of suction line of rocker sprayer is:

(i)8 m   (ii)5 m   (iii)4 m   (iv)10 m

Ans: (ii)8 m

26. Spray droplet size for plant surface application is:

(i) 60- 150 micron,   (ii) 40-50 micron,   (iii)80 micron,   (iv) none

Ans: (i) 60-150 micro

27. Weight carrying capacity of hand rotary duster is –

(i) 2.5-5.0kg   (ii) 4-5 kg   iii) 2-2.5kg

Ans: 4-5 kg

28. The spray is classified as mist when average droplet size is:

a) 50 to 100 µ   b) 0.1 to 50 µ   c) 100 to 200 µ d) 200-400 µ

Ans: Option A (50 to 100 µ)

29. Swirl plate is a part of –

a) Pump   b) Spray lance   c) Nozzle   d) Agitator

Ans: Option c (Nozzle)

30. Which is also called as the "Kitchen Sprayer" –

a) Knapsack sprayer   b) Foot sprayer   c) Bucket pump sprayer   d) Hand sprayer

Ans: Option d (Hand Sprayer)

31. Regular fan type nozzle is used for –

a) Band spraying   b) Broadcast Spraying
c) Both   d) none of these

Ans: Option b (Broadcast spraying)

32. Electrodyne sprayer is a

a) Conserved Droplet Applicator   b) Controlled Droplet Applicator
c) Controlled Diameter Applicator   d) Limited Droplet Applicator

Ans: Option b (Controlled Droplet Applicator)

33. Pressure developed in pedal sprayer is?

a) 12-16kg/cm sq   b)16-20kg/cm sq
c)17-21kg/cm sq   d) 13-7kg/cm sq

Ans: 17-21 kg/cm sq

34. The capacity of knapsack sprayer is?

a) 10-20 L   b) 20-30 L
c)30-40 L   d) 40-50 L

Ans: 10-20 L

35. What is the color code for coarse droplet size?

a) green   b) red
c) blue   d) white

Ans: Blue

True or False

1. Sprayer is a machine which is used to apply pesticides in the form of dust while duster is applied in the form of droplets

Ans: False

2. Dusting is less efficient than spraying because of the low retention of dust.

Answer: True

3. The foot sprayer is an all-purpose sprayer suitable for both small- and large-scale spraying

Answer: True

4. When several nozzles are fitted in a pipe it is called a spray gun

Answer: False

5. Low volume spray has more coverage and are more efficient than high volume spray

Answer: True

6. Tractor mounted sprayer use hydraulic energy.

Answer: True

7. Knapsack power sprayer uses gasoline as fuel.

Answer: True

8. The capacity of hand sprayer is 5 L.

Answer: False (0.5L-3L)

9. Power sprayers can be used for trees up to a height of 15 m

Answer: - True

10. Droplet size for flying insect is 100 microns.

Answer: False (10 – 15 microns is the droplet size)

11. Tank capacity of Aeroblast sprayer is 230 L.

Answer: False(400 L is the tank size)

12. Example of a high-volume sprayer is LV sprayer.

Answer: False

13. rpm of high volume sprayer is 4000-5000.

Answer: False

14. Bucket pump sprayer has a built-in tank for storing the spray fluid

Answer: True

15. The common principle involved in duster operation is to utilize the air blast or air stream

Answer: True

16. The filter of the sprayer protects the pump from abrasion or corrosion

Answer: True

17. Rocker sprayer is a type of pneumatic sprayer

Answer: False

18. Fogging machines are used for spraying the flying small insects

Answer: True

18. Operating pressure below 1.5kg/cm sq is undesirable in nozzle because it does not work properly.

Answer: True

19. The fan nozzle forms a circular spray pattern.

Answer: False, The fan nozzle forms an elliptical spray pattern.

20. The color code of droplet size for very fine category particles is white.

Answer: False, The color code of droplet size for very fine category particle is red.

21. In high volume spray, the dilute liquids are applied by hydraulic machines.

Answer: True

Fill in the blanks

1. Ultra-low volume sprayer's droplet size varies between _____
Ans: 35-100 micron.

3. A prime mover is needed to supply _____ to the power sprayer
Ans: power

4. A _____ is there at the bottom of the tank for draining the liquid.

Ana: drain plug

5. The pressure developed in knapsack hand compression sprayer varies from _____.

Ans: 3 to 12 kg/cm sq

6. A micron is _____ of a millimeter.

Ans: 1/1000

SAQs

1. What is a sprayer? What are the main functions and purpose of a sprayer?

Sprayer is a machine to apply fluids in the form of droplets. Sprayer is used for the following purpose.

- Application of herbicides to remove weeds

- Application of fungicides to minimize fungal diseases

- Application of insecticides to control insect pests

- Application of micro nutrients on the plants

The main functions of sprayer are: -

- To break the liquid droplets of effective size.

- To distribute them uniformly over the plants

- To regulate the amount of liquid to avoid excessive application.

2. What are the desirable qualities of a sprayer?

» It should produce a steady stream of spray material in desired droplet size so that the plants to be treated may be covered uniformly.

» It should deliver the liquid at sufficient pressure so that the spray solution reaches all the foliage and spreads uniformly over the plant body.

» It should be light in weight yet sufficiently strong, easily workable and repairable.

3. Explain about different types of nozzles.

(A) Hollow Cone Nozzle:

This liquid is fed into a whirl chamber through a tangential entry or through a fixed spiral passage to give a rotating motion. The liquid comes out in the form of a harrow conical sheet which then breaks up into small drops.

(B) Solid Cone Nozzle:

This nozzle covers the entire area at small range. The construction is similar to hollow cone nozzle with the addition of an internal jet which strikes the rotating liquid just within the orifice of discharge. The breaking of drop is mainly due to impact.

(C) Fan Nozzle:

It is a nozzle which forms narrow elliptical spray pattern. In this type the liquid is forced to come out as a flat fan shaped sheet which is then broken into droplets. This nozzle is mostly used for low pressure spraying.

4. How is a sprayer classified based on volume of spray?

High volume (HV) sprays: High volume sprayers usually produce a wide range of different size of droplets (>100µm in diameter, usually 300-400µm).

Low volume (LV) sprays: It produces very small droplets (70-150 micron in diameter). Low volume sprayers usually apply < 400 L of spray fluid per hectare

Ultra-low volume (ULV) sprays: The insecticide applied undiluted, in small quantities, usually @ 0.5-6L/ha with ULV sprayer or centrifugal sprayer. The size of the droplet varies from 20-100 micron

5. Why are larger droplets unsuitable for spraying?

Large droplets are undesirable because they often, (a) runoff of the spray target, (b) tend to pool in cupped surfaces, and on lower edges of plants where the higher dose risks phytotoxicity, (c) often do not provide good coverage on the underside of the leaves, (d) require a larger volume of spray than smaller droplets to cover the surfaces, resulting in relatively high amount of active ingredient requirement per unit area.

6. Writes the advantages and limitations of low volume sprays.

Advantages of LV sprays:

> More area coverage in less time

> Less cost (in transport of water, and less labour)

> Timely control of pest is possible due to speedy spray application

> No hazards in handling, as the original container can be directly attached

> Weight of spray appliance is very less

> Effective against epidemics, as the larger area can be sprayed in short time

> More effective since the concentration is high

Disadvantages of LV sprays:

> Risk of loss of chemical due to evaporation of water

> Risk of loss of chemical due to drift

> Specialized formulations are required

7. What are components of a hand compression sprayer?

The typical hand compression sprayer comprises a tank for holding spray material and compressed air, vertical air pump with a handle, filling port, spray lance with nozzle and release and shut-off devices. Besides, it has a metal or plastic skirt which protects the bottom of the tank of the sprayer against wear and makes the sprayer stable when placed on the ground. It also serves as a base for the back- rest. In addition, it has adjustable straps. These should be made of cotton belt, leather on plastic.

8. Where is rocker sprayer used? What are the parts of this sprayer?

For spraying on tall trees like coconut, areca nut, sugarcane, rubber plantations, orchards, vineyards and field crops, vegetable gardens, flower crops etc

This sprayer consists of pump assembly, platform with frame and fork, operating lever, pressure chamber, suction hose with strainer, delivery hose, extension rod with spray nozzles, etc. There is no built-in tank and separate spray tank is necessary.

9. Write a short note on motorised knapsack sprayer.

Knapsack motorized sprayer are the versatile and simple power operated machines. The spray liquid is flown out by means of an air current generated in the machine. They deliver 6.8 to 42.5 m3 (240 to 1500 ft3) of air per minute at a velocity of 200-420 km (125- 260 miles) per hour at the nozzle. The tank, which has a capacity of 10-12 lit, is mostly made of high-density polyethylene. Another small tank of 10-15 lit capacity is provided for the fuel. Generally, they are powered by 1.2 – 3.0 hp petrol engines and the frame is provided with shock-proof cushion which comfortably fix on the back of the operator to eliminate vibrations of the engine. The delivery hoses are very small. It is used for spraying in orchards, coffee estates and tall crops.

10. What are the components of a foot or pedal sprayer?

The foot or pedal sprayers, as they are commonly called, consist of plunger assembly, stand, suction hose, delivery hose, extension rod with a spray nozzle etc. One end of the suction hose is fitted with strainer and the other with a flexible coupling. Similarly, the delivery hose has one end fitted with a sheet off pistol and the other with a flexible coupling. Foot instead of hand operates it, but the principle is the same as in case of the rocker sprayer. This sprayer also does not have a built-in tank. Constant pedalling is required for continuous spray.

11. What is the use of tree sprayer?

Tree sprayer is used to apply pesticide to tall fruit trees.

12. What is the lance length of rocket sprayer?

Lance length of rocket sprayer is usually 60-90 cm.

13. What kind of sprayer is suitable for vineyard?

Blower sprayer is exclusively used in vineyards.

14. Name the parts of a foot sprayer.

The sprayer consists of a pump operated by the foot lever, suction hose with strainer, delivery hose, spray lance fitted with shut off pistol valve, gooseneck bend and adjustable nozzles.

15. State various uses of sprayers.

Spraying is employed for a variety of purposes such as application of: i. Herbicides in order to reduce competition from weeds, ii. Protective fungicides to minimize the effects of fungal diseases, iii. Insecticides to control various kinds of insects pests, iv. Micro-nutrients such as manganese or boron.

16. State the droplet size for soil application.

The droplet size for soil application is 250 – 500 microns.

17. What are the uses of rocker sprayer?

For spraying on tall trees like coconut, sugarcane, rubber plantations, orchards, vineyards and field crops, vegetable gardens, flower crops etc.

18. What is the most commonly used type of hand compression sprayer?

Non pressure retaining type is the most commonly used hand compression sprayer

19. What does low volume sprayer use as carrier of pesticide?

Air stream from a fan.

20. Which duster contains piston?

Plunger duster

21. How much area can a plunger duster cover?

0.5-1.0 ha/day

22. How the spray lance is modified for spraying under the surface of the leaves?

The spray lance is modified by making a bent of 1200 to form a goose neck for spraying under the surface of the leaves.

23. What are the cut off valves used for in a sprayer?

Cut off valves are used to shut off the liquid in a sprayer.

24. What do you mean by VMD?

The size of spray droplet is expressed as volume median diameter (VMD). In other words, 50% of the volume is composed of droplets smaller than the VMD and 50% of the volume is in larger droplets.

25. When the flame thrower is used?

Flame thrower is used to destroy insects like locust swarm and hairy caterpillars, noxious weeds which are to be removed by burning.

26. When ULV application is preferred?

ULV application is preferred when large areas are to be treated.

27. What is the advantage of flood jet nozzle?

Flood jet nozzle is advantageous for minimizing the drift of chemicals.

28. Why a part of manually operated compression sprayer container is kept empty?

A part of the container is kept empty so that adequate air pressure can be developed over the spray fluid in the tank.

29. What is the basic function of a sprayer?

The function of a sprayer is to atomize the spray fluid into small droplets and eject it with some force.

30. What are the two types of nozzles available for power sprayer?

Two types of nozzles are available for power sprayer: jet nozzle and adjustable micronizer nozzle.

31. What are the three types of cut-off valves?

Three types of cut-off valves are -Wheel cut- off valve with strainer; Trigger cut off valve with strainer; Trigger cut off valve without strainer

32. What are the components of a power sprayer?

The components of a power sprayer are as follows –

(i) Prime mover - Prime mover is needed to supply power to the power sprayer. It is usually combustion engine. The power generally varies from 1 to 5 HP.

(ii) Tank - Steel tank is widely used to prevent corrosion. Plastic tanks are also getting popular due to freedom from corrosion and ease of moulding into smooth shapes. A covered opening, fitted with a removable strainer is provided for easy filling, inspection and cleaning. A drain plug is there at the bottom of the tank for draining the liquid.

(iii) Agitator - Agitator are needed to agitate the liquid of the tank. Propeller or paddle type mechanical agitators are provided for agitating the liquid. Horizontal shaft may be used with flat blades rotating at about 100 to 120 rev/min. paddle tip seeds in excess of 2.5 m/sec may cause foaming.

(iv) Air chamber - An air chamber is provided on the discharge line of the pump to level out the pulsations of the pump thereby providing a constant nozzle pressure.

(v) Pressure gauge - The pressure gauge is provided on the discharge line to guide the operator regarding spray pressure. The spray pressure should be under specified limit.

(vi) Pressure regulator - It is meant for adjusting the pressure of the sprayer according to the requirement of the crops in the field.

(vii) Strainer - A strainer is included in the suction line between the tank and the pump to remove dust, dirt and other foreign materials.

(viii) Boom - Field sprayer to be driven by a tractor has a long boom in a horizontal place on which nozzles are fixed at specified spacing. The boom can be adjusted vertically to suit the height of plants in different fields.

(ix) Nozzle - It is used to break the liquid into the desired spray and deliver it to plants. A nozzle consists of: (a) body (b) screw cap (c) disc (d) washer (e) vortex plate (f) strainer.

Usually, the flow rate for a particular nozzle is proportional to the square root the pressure and the discharge rate is proportional to the orifice area. Nozzles have smaller angles. Operating pressure below 1.5 kg/cm2 is undesirable because the nozzle does not work satisfactory.

33. Write a short note on motorised knapsack duster?

Knapsack dusters are common in India. The capacity of the hopper is about 9 kg. The discharge of the dust can be controlled by rotating the plated hose on the blower elbow, which carries the discharge hose. In the spraying cum dusting unit, the conversion of sprayers to duster is very simple and effected by replacing the liquid feed tubes by appropriate dust feeds. Agitation is provided by directing a part of air from the discharge. The dust tank may be the same as for the spray unit except that all the liquid feeds are removed. In some cases, there is a separate unit for replacing the spray lance. Part required for conversion of a sprayer into duster can be purchased at nominal cost from the suppliers of knapsack sprayers cum dusters.

Chapter - 12

Compatibility of Pesticides and Phytotoxicity of Insecticides

Compatibility of Pesticides

Compatibility is the ability of two or more components of a pesticide mixture to be used in combinations without impairment of toxicity, physical properties or plant safety of either of the component.

Cox (1941) coined the term 'Compatibility'. In prophylactic pest control treatment, two or more pesticides, fungicides or even fertilizers are sprayed or applied in the same operation to minimise cost of labour.

Some important criteria of compatibility are:

1. They react adversely to decrease in the activity of the pesticides or chemicals. 2. They do not reach on resultant products that cause injury to crop 3. They do not produce explosive reaction. In compatibility may be due to the following.

Even though guidelines have been presented with respect to tank mixes there still remains the question of compatibility when mixing two or more chemicals, especially, when directions for mixing and application are not included on the label. Both chemical and physical incompatibility is possible. With chemical incompatibility the chemical may be completely deactivated, resulting in no weed control or the chemical might be made highly phytotoxic resulting in damage to the crop. It is also possible to change the mammalian toxicity making a normally safe chemical highly toxic. Physical incompatibility is most commonly evidenced by precipitation in the spray solution which takes the form of crystalline solids, formation of a gelatinous mass, or separation of components which takes the form of layering. Lack of compatibility may only result in the formation of a substance that plugs up screens and nozzles, however, extreme incompatibility may produce a settling out of material that can harden like concrete in the bottom of a tank and in hoses, pumps, and other internal parts of the sprayer. The result may be total loss of the pesticide and use of the sprayer.

Four types of Interaction

- » Additive effects
- » Synergistic effects
- » Antagonism
- » Enhancement

Additive effects

- » Same response when used alone
- » Ease of mixing
- » Reduces application time and labour
  Eg. Root absorbed herbicide with a foliar absorbed herbicide

Synergistic response

- » Confused with additive effects
- » Greater response when mixed
- » True positive interaction among chemicals
- » Reduce cost, time etc.
  Eg. Piperonyl Butoxide and Pyrethrum

Antagonism

- » Less control when 2 or more chemicals are mixed
- » May cause phytotoxicity
  Eg. Mixing of grass and broad leaf herbicide

Enhancement

» When a pesticide is mixed with an additive to provide greater response Eg. Adjuvants

Types of Incompatibility

1. Chemical incompatibilities are more common than physical incompatibilities. Chemical incompatibilities occur when the chemical reaction of the spray mixture increases the effectiveness of one or both materials, or when the chemicals react to form a precipitate. The results may be additive or synergistic (enhancing each other). Or, the action may be negative. Here, the deactivation of a.i. of pesticide is due to pH, temperature, pesticide chemistry etc.

2. Physical incompatibilities involve the inert ingredients to become unstable, forming crystal, flakes, precipitate down of pesticides; it can result in scrumming, lumping, and foaming and can damage equipment. Physical incompatibilities may result when: a) mixing a granular formulation in water, or (2) mixing together wettable powders, oils, and water.
   Example - EC formulation with WP

3. Mechanical Incompatibility: Different pesticides may require different droplet size to be most effective due to variation in spray volumes, adjuvant recommendations besides the over agitation may cause foaming in some older products. In case of soybeans smaller droplet size required for fungicides. But larger droplets to avoid herbicide drift will negatively affect fungicide efficacy.

4. Phytotoxic Incompatibility (Biological incompatibility): When two or more pesticides used in combination result in injury to the host plants. Symptoms include chlorotic spots, darkened shallow pits on fruits, scorching and bleaching of foliage and reduced growth. They often not caused by active ingredient but by adjuvant. It is common in oil-based pesticide mixtures (E.C. formulations). Mixture of sencor (metribuzin) and Tilt (propiconazole) was incompatible and highly phytotoxic when sprayed against foliar diseases of wheat i.e, rust, powdery mildew and blight.

General principles for mixing pesticides:

1. Mix only two pesticides whenever possible and be certain of compatibility. Check the pesticide label for tank mix recommendations and note any restraints.

2. No label recommendations exist, do a jar test prior to spraying. Mix all pesticides properly and according to labels. Make sure all components of spraying have the correct filters. If there is a spray blockage, try to retrieve the mix, before disposing of the tank mix.

3. Test pH. Many incompatibilities result from excessively alkaline (sometimes acidic) pH in the tank. The addition of buffering adjuvants can help.

4. Make a test application to expose any phytotoxicity or antagonism before you make a large-scale application. If you overlap a few strips, this also can show you how much of a margin of safety you have. Wait a few days for symptoms to become visible.

5. Do not mix two emulsifiable concentrates (ECs) (two solvents). This often results in phytotoxicity, the deposit may be decreased due to excess runoff.

6. Do not mix two wettable powders (WPs) unless you know differently. It is commonly done and recommended. For example: benomyl WP and mancozeb WP. It is here not so much as an incompatibility problem, but sometimes a heavy and unsightly residue on the foliage is left. Flower producers are concerned with this as a marketing problem.

7. Do not mix strongly alkaline pesticides with acidic materials

8. Use of EC and WP formulations often results in phytotoxicity due to carriers, emulsifiers and solvents hence take care.

9. Don't mix granular formulations with liquids

10. Use formulations from same manufacturers for mixing if possible. Mixing anionic surfactant pesticides with cationic surfactant pesticides forms precipitate, so use non ionic surfactant pesticides.

Agrochemicals combinations and their significance

Fertilizer-herbicide combinations are extremely popular because they combine two operations. Combinations with pre emergence chemicals are generally effective since both fertilizer and herbicide action are dependent on contact with the soil (requiring rainfall or irrigation). Post emergence herbicide action depends more on absorption by leaves, and granules in such combinations do not adhere well to smooth-surfaced leaves. They will stick better if applied when weed leaves are damp, perhaps with morning dew. "Weed and feed" materials present a conflict in desirable actions. Proper time for weed control often does not coincide with the most desirable time and rates for fertilizing. If used for followup fertilizations, there is danger of herbicide overdose.

Herbicide - Insecticide Combinations

Emulsifiable concentrate formulations of insecticides can act like oil adjuvants when applied in combination with a herbicide.

Test Procedure for physical incompatibility

1. In a clean glass jar with a tight, sealable lid add approximately 50% of the final volume of carrier (water or liquid fertilizer if this is to be used as the carrier) e.g.1L of total solution is to be made up then add 250 mL to the jar at this stage. Note:Use the same source of water that will be used for the tank mix and conduct the test at the temperature the tank mix will be applied.

2. Add a compatibility agent if one is to be used (e.g. ammonium sulphate, Supa Link TM). Shake or stir gently to mix.

3. Next add the appropriate amount of pesticide(s) in their relative proportions based on recommended label rates (refer to the table below). If more than one pesticide is used, add them separately with dry pesticides first, flowables next, and emulsifiable concentrates last. After each addition, shake or stir gently to thoroughly mix.

4. After the pesticides are mixed, add any adjuvants that are to be used in the final spray solution at their recommended label rates.

5. Fill the jar to the required final volume

6. Put lid on and tighten. Invert the jar ten times to mix. Let the mixture stand 15-30 minutes and then look for separation, large flakes, precipitates, heat, gels, heavy oily film on the jar, or other signs of incompatibility.

7. If the mixture separates, but can be remixed readily, the mixture should be able to be sprayed as long as good agitation is used.

8. If the mixtures are incompatible, test the following methods of improving compatibility: (a) slurry the dry pesticide(s) in water before addition, (b) add a compatibility agent if one was not used or (c) add ½ the compatibility agent to the fertilizer or water and the other ½ to the emulsifiable concentrate or flowable

pesticide before addition to the mixture. If incompatibility is still observed, do not use the mixture.

9. After compatibility testing is complete, dispose of any pesticide waste in accordance with the storage and disposal section on the label.

Specific mixtures to avoid

1. Fixed copper fungicides are not generally compatible with insecticides.

2. Roundup herbicide and residue soil applied herbicides result in decreased activity of Roundup.

3. EPTC followed by MCPA or 2,4-D or dinoseb could cause greater injury to desirable plants.

4. Captan, captafol (Difolotan), and folpet (Phaltan) and oil are phytotoxic even when sprayed on the plant separately but within 30 days.

5. Captan and dinitros, fixed coppers, and oils result in phytotoxicity.

6. Acti-dione should not be mixed with highly alkaline materials.

7. Do not use (dimethoate) Cygon or (Captan) Orthocide with oil--phytotoxicity results.

8. Benomyl decomposes in alkaline solutions so should not be mixed together, but can be mixed with fixed coppers if pH is not over 8.0.

9. Do not use Kelthane or Sevin with alkaline materials.

10. Do not combine Meta Systox-R with lime sulfur or fixed copper or Bordeaux. Meta Systox-R is compatible with most other pesticides.

Tests of compatibility

Combinations may either prove phytotonic or phytotoxic sometimes Physical and chemical tests undertaken for testing of insecticide quality and formulations

1. Acidity and alkalinity test

2. Emulsion stability test

3. Wettability test

4. Sieve test

5. Bulk density test

6. Suspensability test

Jar Test for Compatibility

The jar test may be used to test the compatibility of herbicides with each other or herbicides and other pesticides with liquid fertilizers.

1. Add 1 pint of carrier (water, liquid fertilizer) each to two quart jars. Mark the jars with an identifiable letter, number or other means. Usually "with" and "without" is the most practical (representing with and without compatibility agent).

2. Add 1/4 teaspoon or 1.2 ml of compatibility agent to one jar (equivalent to 2 pints per 100 gallons of spray solution).

3. To each jar add the required amount of pesticide in the order suggested in the section on mixing herbicides Shake well after each pesticide addition to simulate continuous agitation.

4. When all ingredients are added, shake both jars for 15 seconds and let stand for 30 minutes or longer. Then inspect the mixture for flakes, sludge, gels, or non dispersible oils, all of which may indicate incompatibility. If, after standing 30 minutes, the components in the jar with no compatibility agent are dispersed, the herbicides are compatible and no compatibility agent is needed. If the components are dispersed only in the jar containing the compatibility agent, the herbicide is compatible only if a compatibility agent is added. If the components are not dispersed in either jar, the herbicide-carrier mixture is not compatible and should not be used.

Phytotoxicity of insecticide

»   Plant Damage due to application of pesticides to plants is known as phytotoxicity.

»   Phyto toxic means harmful of lethal to plant.

»   Plant Damage due to application of pesticides to plants is known as phytotoxicity.

»   Pesticide phytotoxicity appears in several ways on ornamental plants, but probably 5 types of damage most commonly occur.

Causes of Phytotoxicity

Phytotoxicity can occur when:

1. A material is properly applied directly to the plant during adverse environmental conditions.

2. A material is applied improperly.

3. A spray, dust, or vapor drifts from the target crop to a sensitive crop.

4. A runoff carries a chemical to a sensitive crop.

5. Persistent residues accumulate in the soil or on the plant.

Factors which influence phytotoxicity

Chemical - Certain crops are very sensitive to certain chemicals. Be certain the crop to be treated is listed on the chemical label. Some chemicals are persistent. Repeated applications result in accumulation of the chemical to a toxic level.

Formulation - Dusts and wettable powders are generally less phytotoxic than emulsifiable concentrates (EC).

Additives (adjuvants) such as spreaders, stickers, and wetting agents may cause injury.

Concentrations - The use of a chemical concentration higher than label recommendation or its use more frequently than the label recommends is likely to cause plant injury and is illegal.

Method of application - Always use the method recommended by the label. Apply the chemical thoroughly and evenly. Spray dripped on the ends of rows or benches when slowing down to begin the next sweep and excessive overlapping results in some plants receiving to much chemical. High pressure sprays may force chemical into sensitive tissues.

Growing conditions

Temperatures during and after treatments should be moderate. High temperatures favor chlorinated hydrocarbon and sulfur toxicity. Low temperatures favor oil, carbamate, and organophosphate toxicity.

Humidity or plant wetness. Wet foliage at the time of application or prolonged wetness of foliage after spraying can result in injury.

Growth stage of plants. Seedlings and fast-growing, succulent plants are usually sensitive to chemical treatment.

Mixing incompatible chemicals. This can occur when two materials are deliberately applied as a mix or if a materialis applied too soon after a previous material was used.

Symptoms

Symptoms of phytotoxic reactions. Plants react to pesticides in a number of ways:

1. Chlorosis – appears as spots or as tip, margin or leaf yellowing.

2. Leaf distortion – appears as curling, crinkling or cupping of the leaf.

3. Stunting – The entire plant is reduced in size or certain parts (fruit, flowers, roots) are smaller while the rest of the plant appears normal.

4. Abnormal growth – Excessive growth on either certain parts (aerial roots, suckering) or the entire plant.

5. Pesticide phytotoxicity and other types of chemical injury, including damage from air pollution and fertilizers, can often be distinguished from biological pest problems by the pattern and timing of symptom development.

6. Death of seedlings.

7. Death of rapidly growing succulent tissues Stunting or delayed plant development.

8. Russeting or bronzing of leaves or fruit.

9. Dead spots or flecks on leaves.

10. Dead leaf tips or leaf margins.

11. Dead areas between the veins of the leaves

12. No signs of plant pathogenic organisms.

13. Injured leaf tissue is sharply defined with little or no colour gradation from dead areas into healthy areas.

14. Dead spots are of uniform colour and may go entirely through the leaf.

15. Injury occurs over a relatively short period and does not spread from plant to plant.

16. Only tissue of a certain age may show damage (only young leaves).

17. Plants on ends of rows or ends of benches are the primary ones affected.

Control or reduce management

The following are general rules or guidelines to help reduce phytotoxicity

1. Don't apply a pesticide to plants that are stressed. Plants should be growing at their optimum.

2. Avoid spraying under extremely hot, sunny conditions. Spray in the mornings when possible.

3. Preferably between 6 and 10 a.m. When air or plant tissue temperature is approximately 90°F or higher, damage will likely occur. On bright sunny days, leaf tissue temperatures may be higher than the surrounding air, thus increasing the possibility of injury. Also, slow growing plants due to cool weather or other conditions (i.e. overcast, low light conditions) are more likely to be damaged. Avoid temperature extremes, either high or low.

4. Don't apply pesticides under conditions which will not promote drying. Plants sprayed when cool, humid conditions exist for extended periods will remain wet for long periods of time and increase the probability of injury. This is one of the reasons plants sprayed under greenhouse conditions are more likely to be damaged.

5. However, never spray plants when they are in need of water. Wilted or dry plants are extremely sensitive to spray injury.

Practical exercise on phytotoxicity

Objective:

To evaluate phytptoxicity of Prahar-C (Profenophos 40 % + Cypermethrin 4 % EC).

Crop: Brinjal

Treatment details

S. no.	Treatment	Dosage (ml/ha)	Application time & water volume/ha
1	Untreated control (water spray)	–	Give one spray as soon as the shoot and fruit borer and sucking pest complex reaches ETL. Water volume of 400 – 500 litres/ha to be used by using hollow cone nozzle depending upon the stage of the crop at spray time.
2	Prahar-C (Profenophos 40 % + Cypermethrin 4 % EC).	1500	
3	Prahar-C (Profenophos 40 % + Cypermethrin 4 % EC).	3000	
4	Prahar-C (Profenophos 40 % + Cypermethrin 4 % EC).	6000	

Methodology: Take observations on plants per treatment. Count total number of leaves and leaves showing phytotoxic symptoms if any and workout the phyto-toxicity percentage and apply phytotoxicity scale (0-10) as given below under point no. 8.

Observations: Phytotoxicity scale (No phytotioxicity = 0 scale; 1-10 % phytotoxicity = 1 scale; 11-20 % phytotoxicity = 2 scale; 21-30 % phytotoxicity = 3 scale; 31-40 % phytotoxicity = 4 scale; 41-50 % phytotoxicity = 5 scale; 51-60 % phytotoxicity = 6 scale; 61-70 % phytotoxicity = 7 scale; 71-80 % phytotoxicity = 8 scale, 81-90 % phytotoxicity = 9 scale; 91-100 % phytotoxicity = 10 scale) to be observed on different parameters (leaf chlorosis, leaf tip burning, leaf necrosis, leaf epinasty, leaf hyponasty, vein clearing, wilting &rosetting) at 1 & 5 days after spray.

Different phytotoxic parameters

Leaf tip burning

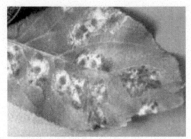

Leaf necrosis

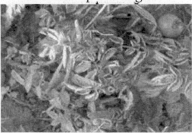

Leaf hyponasty

Wilting

Leaf epinasty

Leaf vein clearing

Rossetting

Chlorsis

SAQs

1. What do you mean by Pesticides?

»   Pesticides are chemical substances that are meant to kill pests.

2. What is Agrochemical?

»   An agrochemical is any substance that humans use to help in the management of an agricultural ecosystem.

3. Who coined the term 'Compatibility'?

»   Cox (1941) coined the term 'Compatibility'.

4. What are the basic criteria of compatibility of pesticides?

»   They react adversely to decrease in the activity of the pesticides or chemicals.

»   They do not reach on resultant product that cause injury to crop.

»   They do not produce explosive reaction. In compatibility may be due to the following.

5. What is phytotoxicity?

»   Plant Damage due to application of pesticides to plants is known as phytotoxicity.

6. What is Chlorosis?

»   A yellowing or bleaching effect of plant.

Chapter - 13

Study of Insect Pollinators, Weed Killers and Scavengers

Insect Pollinators

Pollination refers to the transfer of anther to stigma in flowering plants for sexual reproduction. Insects aid in cross pollination in fruits, vegetables, ornamentals, cotton, tobacco, sunflower and many other crops.

Role: Insect pollination helps in uniform seed set, improvement in quality and increase in crop yield.

Entomology refers to cross pollinators aided by insects

Pollination class	Type of insect
Melitophily	Bees
Cantharophily	Beetles
Myophily	Syrphid and Bombyilid flies
Sphigophily	Hawk moths
Psycophily	Butterflies
Phalacophily	Small moths

1. Honeybees as pollinators

All bee species aid in pollination Value of honey bees in pollination is 15-20 times higher than that of the honey and wax it produces. Per cent increase in yield due to bee pollination in Mustard - 3%, Sunflower - 32 - 48%, Cotton – 17 - 19%, Lucerne - 112%, Onion - 93%, Apple - 44%, Cardamom - 21-37% etc.

2. Hoverflies (*Syrphus sp*) (Syrphidae : Dipteria)

Brightly coloured flies body is striped or aided with yellow or blue resemble bees and wasps, larval stage predatory, adults are pollinators

Crops pollinated - carrot, cotton, pulses.

3. Carpenter bee, *Xylocopa sp*. (Xylocopinae:Anthophoridae):

Robust dark bluish bees with hairy body, dorsum of abdomen bare, pollen basket absent, adults are good pollinators, construct galleries in wood and store honey and pollen.

4. Digger bees, (*Anthophora sp*.) (Anthophoridae:Hymenoptera):

Stout, hairy, pollen collecting bees ans abdomen with black and blue bands

5. Fig wasp, *Blastophaga psenes* (Agaonitae:Hymenoptera)

Fig is pollinated by fig wasp only. There is no other mode of pollination. There are two types of fig - Caprifig and Symrna fig.

i.   Caprifig :- a. It is a wild type of fig - not edible b. Has both male and female flowers c. Pollen is produced in plenty d. Natural host of fig wasp

ii.   Smyrnafig :- a. It is the cultivated type of fig - Edible b. It has only female flowers c. Pollen not produced d. Not the natural host of fig wasp, Fig wasp: Male - wingless, present in caprifig, Female – winged

Wasp lays eggs in caprifig, larvae develops in galls in the base of the flowers, mates with female even when it is inside gall. Mated wasp emerges out of flower (caprifig) with lot of pollen dusted around its body.

The fig wasp enters smyrna fig with lot of pollen and deposits it on the stigma. But it cannot oviposit in the ovary of symrna fig which is deep seated. It again moves to capri fig for egg laying. In this process smyrna fig is pollinated. Caprifig will be planted next to smyrna fig to aid in pollination.

6. Oil palm pollinating weevil: *Elacidobins kamerunicus* **(Curculionidae: Coleoptera):** Aid in increasing oil palm bunch weight by 35% and oil content by 20%.

7. Other pollinators:

Butterflies (eg *Deilaphila spp.*) and moths (*Acherontia spp.*), ants, flies, stingless bees, beetles etc.,

Weed killers:

Insects which help to controlling weeds by feeding on them are called weed killers.

1. *Dactylopius tomentosus* (Cochnieal insect) to control prickly pear *Opuntia dillenii*. This insect was introduced into India in 1925. Within 5-10 years it controlled the weed.

2. Aristalochia butterfly, *Papilio aristolochiae* (Papilionidae: Lepidoptera) - It feeds on *Arista lochia* which is a weed.

3. Calotropis butterfly - *Danaus chrysippus* (Nymphalidae:Lepidoptera) - feeds on *Calotropis.*

4. AK Grosshopper - *Poecilocerus pictus* (Actididae: Orthoptera) - feeds on *Calotropis* and controls it

5. Water hyacinth weevil (*Neochetina eichhorniae* and *N. Bruchi*) - The larvae tunnel and feed inside the petioles. Ten pairs of adults and progeny controls plant growth in 0.58 m^2.

6. Parthenium weed killer, *Zygogramma bicolorata* (Chrysomelidae: Coleoptera) - Adults and grubs feed on leaves and flowers. 2 beetles controls and destroys one plant in 45 days.

A successful weed killers has following qualities

  » Should not be a pest of cultivated plants - at present or in future

  » Effective in damaging and controlling the weed

  » Should be a borer or internal feeder of the weed.

  » Should not be affected by Parasitoids /predators

Scavengers:

Insect which feed on dead and decaying plant and animal matter are called scavengers.

Role:

Remove decomposing material and prevent health hazard convert complex material into simple substances.

a. Rove beetles (Staphylinidae: Coleopteran) adults and larvae feed on decaying matter.

b. Chafer beetles (Scarabaeidae: Coleoptera)

c. Darkling beetles (Tenebrionidae: Coleoptera)

d. Nitidulids (Nitidulidae: Coleoptera)

e. Ants (Hymenoptera)

f. Termites (Isoptera)

g. Muscid flies (Muscidae: Diptera)

Important species of pollinators

Honee bee (*Apis spp.*)	Carpenter bee (*Xylocopa spp.*)
Digger bee (*Anthophorini spp.*)	Hover fly (*Syrphus spp.*)

Fig wasp (*Blastophaga spp.*)	**Elephant hawkmoth (*Deilaphila sp.*)**
Death's head hawkmoth (*Acherontia sp.*)	**Stingless bee (*Meliponini spp.*)**

Important species of weed killers

***Cochnieal* insect (*Dactylopius coccus*)**	***Aristalochia* butterfly**
Parthenium weed killer (*Zygogramma bicolorata*)	**AK Grosshopper** (*Poekilocerus pictus*)

| Water hyacinth weevil (*Neochetina spp.*) | Calotropis butterfly |

Important species of scavanger

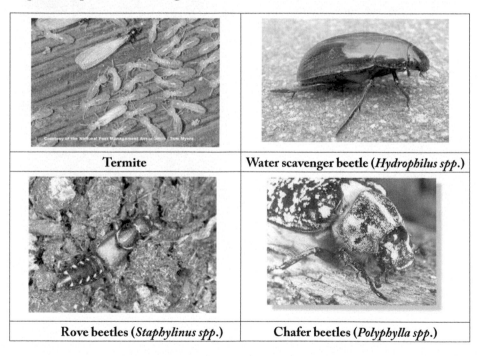

| Termite | Water scavenger beetle (*Hydrophilus spp.*) |
| Rove beetles (*Staphylinus spp.*) | Chafer beetles (*Polyphylla spp.*) |

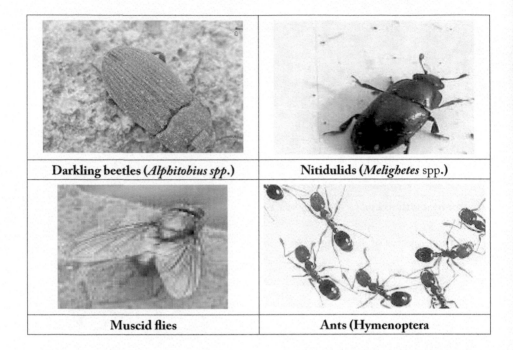

| Darkling beetles (*Alphitobius spp.*) | Nitidulids (*Melighetes* spp.) |
| Muscid flies | Ants (Hymenoptera |

MCQs

1. Anemophily type of pollination is found in –

a. Vallisneria   b. Salvia   c. Coconut   d. Bottle brust

Ans: Coconut

2. Pollination which occurs in closed flower is known as –

a. Allogamy   b. Cleistogamy   c. Protogyny   d. None of the above

Ans: Clestiogamy

3. The exine of a pollen grain is made of –

a. Pectin and Cellulose   b. Sporopollenin   c. Pollenkit   d. Lignocellulose

Ans: Sporopollenin

4. Pollination by bats is called –

a. Ornithophily   b. Entomophily   c. Malacophily   d. Cheriopterophily

Ans: Cheriopterophily

True/False

1. Beetles are weed killers

Ans: True

2. Scavengers are not typically omnivorous

Ans: False

SAQs

1. What is the value of honey bee as pollinator?

Ans: Value of honey bees in pollination is 15-20 times higher than that of the honey and wax it produces.

2. Name a few pollinator species.

Ans: A few pollinator species are: honeybee, carpenter bee, fig wasp, hoverflies, etc.

3. What are types of fig pollinated by fig wasp?

Ans: Caprifig and smyrnafig

4. Name the insect species that acts as a weed killer for prickly pear cactus.

Ans: *Dactylopius tomentosus*

5. What are the qualities of a successful weed killer?

»   Should not be a pest of cultivated plants - at present or in future

»   Effective in damaging and controlling the weed

»   Should be a borer or internal feeder of the weed

»   Should not be affected by parasitoids/predators

6. Who are detritivores?

Ans: Animals that collect small particles of dead organic material of both animal and plant origin are referred to as detritivores.

8. What is called for animals that eat dead animals or carrion?

Ans: Scavengers

9. Who is called scavenger of Earth?

Ans: Fungi are called as scavengers of earth. They decompose waste, so they are called as scavengers of earth.

10. Which is the only obligate vertebrate scavenger?

Ans: Vultures

11. What are weed killers insects?

Ans: Many insects feed upon unwanted weeds, just the same manner they do with cultivated plants. Common example of some weeds killers, prickly pear control with cochineal insect, *Dactylopius tomentosus*

8. What is Scavengers?

Ans: Who eat the flesh of dead animals, as food, are called Scavengers

9. Is cockroach a scavenger?

Ans: Cockroach is omnivorous scavengers that devour keratin. They will bite human flesh in both the living and dead with resultant injury.

10. Which flower is best pollinated by insects?

Ans: Crocus

Chapter - 14

Commonly Used Acaricides, Rodenticides and Nematicides

Acaricides

Acaricides are pesticides that kill members of the Acarina group, which includes ticks and mites. They are also called as miticides (kills mites) and ixodicides (kills ticks). Acaricides are used both in medicine and agriculture.

Group	Example (s)
Macrocyclic lactones	Abamectin, Ivermectin
Oragnochlorines	DDT, Dicofol
Carbamates	Carbaryl, Carbofuron, Propoxur, Aldicarb
Formamidine	Chlordemiform, Amitraz
Organophosphates	Monocrotofos, Chlorpyrifos, Dimethoate, Ethion, Formothion Parathion, Phorate, Phosalone, Triazophos
Phenyl Pyrazole	Fipronil
Pyrethoids	Cypermethrin, fluvalinate
Thiourea derivatives	Diafenthiuron
Pyrroles	Chlorfenapyr
Thiazolidine Group	Hexythiazox
Sulfite Ester Group	Propargite
Ouinazoline Group	Fenazaquin

Tetronic Acid Derivatives	Spiromesifen
Pyridazinones Acaricide	Fenpyroximate

Rodenticides

Rodenticides (Rat Killers): These are chemicals or any other substances intended to kill rodents (rats, mice etc.). Rodents are difficult to kill with poisons due to their uncommon feeding habits. They will eat a small bit of something and wait, and if they don't get sick, then they continue. An effective rodenticide must be tasteless and odorless in lethal concentrations, and have a delayed effect. These are the chemicals to kill mammals, and hence are extremely toxic to humans also. So, it needs to be used under strict supervision of expert. Warfarin and other anti-coagulant chemicals (Coumarins) were initially developed to overcome this problem by creating compounds that were highly toxic to rodents, particularly after repeated exposures, but much less toxic to humans. New super-warfarins (eg: Bromadiolone) are available and toxic at much lower dosages than conventional warfarins. Since the rodents usually share the environment with humans, accidental exposure is an integral part of the placement of baits for the rodents.

Rodenticides		
Inorganic	**Organic**	
Metal phosphides	**Botanical**	**Synthetic**
e.g. Zinc phosphide, Aluminium phosphide, Calcium phosphide, Magnesium phosphide etc.	Strychnine, Red squill etc.	Coumarins/Anticoagulants e.g. Bromadiolone, Warfarins

Metal Phosphides: (Inorganic rodenticides)

Metal phosphides have been used as a means of killing rodents and are considered single-dose fast acting rodenticides (death occurs commonly within 1-3 days after single bait ingestion). Bait consisting of food and a phosphide (usually zinc phosphide) is left where the rodents can eat it. The acid in the digestive system of the rodent reacts with the phosphide to generate the toxic phosphine gas. Phosphides are rather fast acting rat poisons, resulting in the rats dying usually in open areas instead of in the affected buildings.

Zinc phosphide: Baits of zinc phosphide are cheaper than most new generation anticoagulants (super-warfarins). Inversely, the individual rodents that survived anticoagulant bait poisoning (rest population) can be eradicated by pre-baiting them with nontoxic bait for a week or two (this is important to overcome bait shyness and to get rodents used to feeding in specific areas by specific food) and subsequently applying poisoned bait of the same sort as used for pre-baiting until all consumption of the bait ceases (usually within 2-4 days). These methods of alternating rodenticides with different modes of action give actual or almost 100% eradications of the rodent population in the area, if the acceptance/palatability of baits is good. Zinc phosphide is typically added to rodent baits in amount of around 0.75-2%.

Other phosphides used as rodenticides are:

a.   Aluminium phosphide (fumigant only) (Celphos, Quick-phos, Alphos tablets): available in 3 g tablets, widely used in India as grain preservative for storage grain insect pests. Each 3 g tablet liberates 1 g phosphine gas.

b.   Calcium phosphide (fumigant only)

c.   Magnesium phosphide (fumigant only).

Botanical origin Rodenticides

Strychnine: The most common source is from the seeds of *Strychnos nux vomica*.

Red squill: A glycoside as rodenticide is derived from the bulbs of a lily like subtropical plant. It has an emetic-property. So, if rodents ingest a product with red squill, because the rodents are incapable of vomiting, and finally leading to death. It is less toxic to animals other than rodents because it is removed from the stomach by vomiting - a reflex that is absent in rodents.

Synthetic rodenticides:

Coumarins: Warfarins and Bromadiolone.

Coumarins are sweet smelling/pleasantly fragrant chemicals with anti-coagulant activity. Coumarins depress the synthesis of vitamin K. The anti-prothrombin effect is best known. They also increase permeability of capillaries throughout the body, predisposing the animal to widespread internal hemorrhage. This generally occurs after several days of warfarin poisoning / ingestion. These substances kill the rodent by preventing normal blood clotting and causing internal hemorrhage. This is slow acting rodenticides, require pre-baiting.

Warfarin: It is a first-generation coumarin rodenticide that acts as an anti-coagulant of blood by blocking vitamin K cycle, causing Hemophilia. It is now banned for use, as it is highly toxic to humans. It has shorter half-lives, and hence requires high doses /concentrations for consecutive days (multiple dose rodenticides).

Bromadiolone: It is a second-generation coumarin rodenticide with anti-coagulant activity (vitamin K antagonist). They are very toxic to rodents than first-generation coumarins, but due to its usage at very low doses @ 0.001-0.005% in baits (multiple dose rodenticides), they can be used safely. These are also called as super-warfarins since they are lethal at very low doses.

Nematicides: The Substances which are used to control nematodes are called as Nematicides. They are mainly chemical compounds but a very few bio-agents are also available to act as nematicides. Use of chemical nematicides, though a very effective method of nematode control, is generally discouraged due to several reasons. The main consideration is the cost. Nematicides are by and large very costly. Secondly, it is fraught with several side effects like residue problems, resurgence of nematode populations, environmental pollution, human hazards, etc. Nematicides are recommended to be used only when nematode populations are very high, crop is valuable, and quick results are wanted. To make the use of nematicides economically feasible and safer, several ways and means are available.

Nematicides can broadly be grouped into two categories, fumigants and non-fumigants.

Fumigants: The fumigants are generally highly volatile compounds and when applied in the soil turn into gaseous phase. The vapour diffuses through the pore spaces and is toxic to the nematodes. Most of the earlier nematicides were fumigants, involving halogenated hydrocarbons, e.g. DD- a mixture of dichloropropane and dichloropropene, EDB- ethylene dibromide, MBr- methyl bromide, Chloropicrin-trichloro nitro methane, Namagon (DBCP) - dibromo chloro propane. Another group of fumigants which release methyl isothiocyanate, e.g. metham sodium, dazomet etc. was developed later on.

Mode of action: The activity of halogenated hydrocarbon group of fumigants depends upon reaction with nucleophilic site, e.g. OH, SH, NH_2 groups in a vital enzyme system in the nematodes. This site binds with the alkyl portion of the fumigant. Thus, biochemical pathways in protein synthesis and respiration are affected directly. The affected nematode exhibit a period of hyperactivity, followed by paralysis and death.

Nematicidal property of compounds that are precursors of methyl iso thiocyanate is due to the ability to react with the thiol groups in vital enzymes in nematodes. Once inside the nematode, cyanide prevents the utilization of oxygen and so prevents respiration.

Limitations: There are certain problems associated with the use of these fumigants. Highly volatile compounds are phytotoxic and hence necessitate pre-plant application and waitring period of 3-5 weeks. Some of them may need special applicators and plastic covers to prevent the escape of vapours into the atmosphere. DBCP was later found to be carcinogenic and hence banned, besides it leaves bromine residue in the fruits. Moreover, application of such chemicals warrants special field preparations, and their efficacy is largely governed by soil type, soil moisture and temperature, depth of application, etc. Metham bromide has been phased out allegadely for its role in depletion of ozone layer.

Non-fumigants: These nematicides are futher classified according to their chemical groups: carbamates and organophosphates. Names of few commonly used non-fumigant nematicides are given hereunder.

Organophosphates		Carbamates	
Common name	Trade name	Common name	Trade name
Ethoprophos	Mocap	Aldicarb	Temik
Fensulfothion	Dasanit	Carbofuran	Furadon
Phenamiphos	Nemacur	Oxamyl	Vydate
Phorate	Thimet		
Thionazin	Nemaphos		

Mode of action: These nematicides are also known as acetylcholinestarase (AChE) inhibitors as they bind with this enzyme (phosphorylation by organophosphates, carbamoylation by carbamates) and deactivate it. AChE is essential to hydrolyse acetylcholine (chemical transmitter) produced at the synapse during impulse transmission in nervous system. For proper operation of nervous system it is necessary that once the message has been passed, excess acetylcholine is removed from the synapse to prevent repetitive fixing and to allow succeeding message to be transmitted. This removal is effected by AChE.

As a result of inhibition of AChE by organophosphate or carbamate compounds, the liberated acetylcholine accumulates and prevents smooth transmission of nerve impulse across synapse. This results in loss of muscular coordination, convulsions and ultimate death.

Advantages of non fumigants: These are effective against some insects also, much more effective at lower dosages, easy to handle and apply, less phytotoxic and can be applied in standing crop, most of them are systemic and slow release in action and relatively non-volatile.

Application can often be accompanied at the time of sowing, making additional mechanical applications and waiting period after treatment unnecessary.

Econimizing nematicide use:

Most of the nematicides are very costly, hence overall application escalates cost. To cut down the cost of nematicidal application, the following practices can be adopted wherever possible.

a. Seedling bare root dip treatment in systemic nematicides in transplanted crops.

b. Seed treatment through coating with or dipping in nematicides solution.

c. Nursery bed treatment only.

d. Row and spot application for widely spaced crops.

Question - Answer

1. What are acaricides?

Acaricides are pesticides that kill members of the arachnid subclass Acari, which includes ticks and mites.

2. What are ixodicides?

"Ixodicides" are substances that kill ticks.

3. State few examples of acaricides.

Permethrin, Carbamate, Dienochlor, Organophosphate

4. State one example each of pyrethroids type and macrocyclic lactones type of acaricide.

Example of pyrethroids type of acaricides is Permethrin and one macrocyclic lactones type of acaricide is Ivermectin.

5. What is pyrethrum?

"Pyrethrum" refers to the dried and powdered flower heads of *Chrysanthemum cinerariaefolium.*

6. What is pyrethroids?

"Pyrethroids" refer to the insecticides based on the pyrethrins as prototypes.

7. Which compounds open the door for the synthesis of OP analogues?

Schradan and parathion

8. Macrocyclic lactones are isolated from which microorganisms?

Streptomyces avermitilis

9. What are rodenticides?

Rodenticides are chemicals made and sold for the purpose of killing rodents. While commonly referred to as "rat poison", rodenticides are also used to kill mice, squirrels, woodchucks, chipmunks, porcupines, nutria, beavers, and voles.

10. State the 3 modes of poisoning of rodenticides.

Ingestion, inhalation, dermal route

11. State a low toxicity or eco-friendly rodenticides.

Powdered corn cob

12. State a highly toxic rodenticide.

Thallium

13. What is yellow phosphorous?

Yellow or white phosphorous is a waxy substance and is commonly used in fertilizers, fireworks, ammunition, and rodenticide preparations.

14. State few commonly used rodenticides.

Warfarin, 1080 (sodium fluoroacetate), ANTU (legal label for alpha-naphthylthiourea), and red squill are commonly used rodenticides.

15. What is nematicide?

A nematicide is a type of chemical pesticide used to kill plant-parasitic nematodes.

16. Name a tree whose products are used for preparing nematicides.

Neem

17. State few fumigants used as nematicides.

Methyl bromide, Ethylene di-bromide, etc.

18. Why fumigants and non-fumigant nematicides are prohibited in some countries?

Because they causes groundwater pollution.

19. What are ioxicides?

Acaricides used to kill ticks are called ioxicides.

20. Classify nematicides based on volatility.

Fumigants and non- fumigants. Non – fumigants are further divided into two groups
----- Organophosphate and carbamates.

21. How anticoagulants work?

Anticoagulants stop the function of enzyme responsible for recycling of vitamin K in the body of rodents. As a result blood clotting agents which need vitamin K are not synthesized and bleeding does not stop.

22. Why did the use of chlordimeform discontinued?

It is discontinued for it's carcinogenic effect.

23. Name one fumigant which cause ozone depletion.

Methyl Bromide.

24. Mention 2 differences between first and second generation anticoagulants.

First generation anticoagulants	Second generation anticoagulants
It requires multiple feeding for its effectiveness.	It requires less feeding (single dose) for its effectiveness.
It is less toxic to non -target species.	It is more toxic to non-target species.

25. Name some acaricidal plants.

Some famous acaricidal plants are *Azadirachta indica, A. Juss, Melaia azedarach, Tateges erecta* etc.

26. Mention mode of action of Formamidines.

Formamidines are octopamine agonists in the nervous system.

27. Name one multiple dose non-anticoagulant.

Cholecalciferol

28. Mention the factors that contribute to the control efficacy by pheromones.

Rate of release of pheromones, application procedure (on the pest or vegetation), solubility of the pheromone in the coupling device (carrier), the required range of attraction, the duration of the activity etc. contribute to the control efficacy by pheromones.

B. State true or false:

1.   Cyperkill is applied by spray method.

False

2.   Carbaryl is an example of organophosphate.

False

3.   Miticides are substances that kills mite.

True

4.   Fluazuron is an IGR which act as acaricides.

True

5.   DDT cannot kill mites.

False

6.   Carbamates are less safer than OPs.

False

7.   Pyrethrums have high photo instability.

True

8.   Pyrethrums are less safer to mammals than SPs.

False

9.   Pyrethrums do not have knockdown effect on insects.

False

10. Compared to organoclorinated, OPs are much less environmentally persistent, more biodegradable, less subject to biomagnifications and usually unstable in presence of sunlight.

True

11. Macrocyclic lactones act as a nerve poison.

True

12. Rodenticides are commonly referred as rat poison.

True

13. Rodents does not show bait shyness.

False

14. First generation rodenticidal anticoagulants generally have shorter elimination half-lives, require higher concentrations.

True

15. Metal phosphides are considered as single dose first acting rodenticides.

True

16. Aluminium phosphide is used as fumigant only.

False

17. Arsenic trioxide do not act as rodenticides.

False

18. Corn meal gluten is an eco-friendly rodenticide.

True

19. LD_{50} value of highly toxic insecticides ranges between 0-50 mg/kg of body weight.

True

20. Phosphides are the most uncommon agent responsible for rodenticide poisoning in India.

False

21. Coumarins come closest to being an ideal rodenticide.

True

22. Red squill is not a commonly used rodenticide.

False

23. Nematicides have tended to be broad-spectrum toxicants possessing high volatility or other properties promoting migration through the soil.

True

24. An environmentally benign garlic-derived polysulfide product is used as nematicides in some countries

True

25. Carbofuran is a fumigant which can not kill namatodes.

False

26. EDB is a liquid fumigant used to control nematodes.

True

27. Temik is an acaricide.

Ans. False (F).

Fill in the blanks:

1. …….................. are substances that kill ticks.
   Ans: Ixodicides

2. Non-pesticide miticides act by causing ………………………
   Ans: desiccation.

3. The principle active ingredient pyrethrum found in flower ranges from …………...
   Ans: 0.7-3.0%.

4. Pyrethrins give rapid ……………., which is a characteristic feature.
   Ans: knock down action

5. ………………………….. are synthetic analogues of the naturally occurring botanical insecticide pyrethrum.
   Ans: Synthetic pyrethroids (SPs)

6.   .... are the largest group of insecticides.
     Ans: OPs

7.   .... is the most common mode of rodenticide poisoning reported.
     Ans: Ingestion

8.   .... rodenticidal anticoagulant agents are far more toxic than first
     generation.
     Ans: Second generation

9.   Magnesium phosphide is used as .... only.
     Ans: fumigant

10. In some countries, fixed three-component rodenticides, i.e.,..........., are used.
     Ans: anticoagulant + antibiotic + vitamin D

11. Coumarins are .... in rats; however, the toxicity is much less in humans.
     Ans: hepatotoxic

12. ...., a carbamate insecticide marketed by Bayer Crop Science, is an
     example of a commonly used commercial nematicide.
     Ans: Aldicarb (Temik)

13. All nematicides are eventually degraded if they remain in the .... where there
     is greatest microbial activity
     Ans: topsoil

14. Chemical name of Mocap is ....
     Ans: ethoprophos .

15. Formulation of Thionazin is ....
     Ans: Granular and emulsifiable liquid.

16. Second generation anticoagulants are also called ....
     Ans: single- dose anticoagulants.

Multiple choice question:

1. Which of the following is an acaricide?

I. Coumaphos   II. Zinc phosphide   III. Neem

Ans: I

2. Pyrethroids is derived from which species?

I. *Derris sp.*   II. *Nicotiana sp.*   III. *Chrysanthemum sp.*

Ans: III

3. Macrocyclic lactone is a _

I. Nerve poison   II. Stomach poison   III. Physical poison

Ans: I

4. Coumarin is a _

I. Acaricide   II. Rodenticide   III. Nematicide

Ans: II

5. Based on toxicity, Red squill is _

I. Highly toxic   II. Moderately toxic   III. Less toxic

Ans: III

6. Fumigants perform best as nematicides in soils that do not have

I. high levels of organic matter

II. medium levels of organic matter

III. low levels of organic matter

Ans: I

7. Secondary risk in both mammals and birds are low in

a. Warfarin   b. Diphacinone   c. Bromadiolone   d. Brodifacoum

Ans. a. Warfarin

Chapter - 15

Microbial Insecticides and IGRS

Introduction

The term microbial control was coined by Steinhaus in 1949. When microbial organisms or their products (toxins) are employed by man for the control for pests it is referred to as microbial control. Crop pest can be controlled by use of its disease causing microorganisms like viruses, bacteria, protozoa, fungi and nematodes.

Microbial insecticides are especially valuable because their toxicity to non target animals and humans is extremely low. Compared to other commonly used insecticides, they are safe for both the pesticide user and consumers of treated crops. Microbial insecticides also are known as biological pathogens and biological control agents. Microbial insecticides are comprised of microscopic living organisms (viruses, bacteria, fungi, protozoa, or nematodes) or the toxins produced by these organisms. They are formulated to be applied as conventional insecticidal sprays, dusts, liquid drenches, liquid concentrates, wettable powders, or granules. Each product's specific properties determine the ways in which it can be used most effectively.

Traits desirable in pathogens for microbial control

1. Suitable strain

Specific strain of a microorganism is effective for controlling an insect species.

2. Virulence

It is the power of the pathogens to invade and produce disease in their hosts.

3. Toxins

It refers to the poisonous nature of the chemicals produced by the pathogens.

4. Persistence

It is the quality which enables a microorganisms to retain its viability and virulence on storage.

Principle groups of pathogens

The microorganisms that have been found suitable for insect control are placed in two groups: ingested microbes (bacteria and viruses) and non ingested microbes (fungi and nematode) which enter by penetrating the integument.

Bacteria (*Bacillus thuringiensis*)

It is a gram positive, soil dwelling, rod shaped and spore forming crystalliferous bacterium, commonly used as pesticides. It was first discovered in 1901 by Japanese biologist S. Ishiwatari. Upon sporulation, it forms crystals of protinaceous insecticides. CRY toxins have specific activities against species of the orders Lepidoptera, Diptera and Coleopteran. Thus it serves as an important reservoir of CRY toxins and cry genes for production of insecticides and GM crops. Most insecticidal strains have been isolated from soil samples. Bacterial insecticides must be eaten to be effective; they are not contact poisons. Insecticidal products comprised of a single *Bacillus* species may be active against an entire order of insects, or they may be effective against only one or a few species. For example, products containing *Bacillus thuringiensis* var. *kurstaki* kill the caterpillar stage of a wide array of butterflies and moths. In contrast, *Bacillus popillae* (milky spore disease) kills Japanese beetle larvae but is not effective against the closely related annual white grubs (masked chafers in the genus Cyclocephala) that commonly infest lawns.

When Bt is ingested by a susceptible insect, the protein toxin is activated by alkaline conditions and enzyme activity in the insect's gut. The toxicity of the activated toxin is dependent on the presence of specific receptor sites on the insect's gut wall. This necessary match between toxin and receptor sites determines the range of insect species killed by each Bt subspecies and isolate. If the activated toxin attaches to receptor sites, it paralyzes and destroys the cells of the insect's gut wall, allowing the gut contents to enter the insect's body cavity and blood stream. Poisoned insects may die quickly from the activity of the toxin or may die within 2 or 3 days from the effects of septicemia (blood poisoning). Although a few days may elapse before the insect dies, it stops feeding (and therefore stops damaging crops) soon after ingesting Bt.

Additional isolates that kill other types of pests have been identified and developed for insecticidal use. The nature of the crystalline protein endotoxin differs among Bt subspecies and isolates, and it is the characteristics of these specific endotoxins that determine what insects will be poisoned by each Bt product

Use of Bt insecticide

Where Bt is applied to plant surfaces or other sites exposed to sunlight, it is deactivated rapidly by direct ultraviolet radiation. To maximize the effectiveness of Bt treatments, sprays should thoroughly cover all plant surfaces, including the undersides of leaves. Treating in the late afternoon or evening can be helpful because the insecticide remains effective on foliage overnight before being inactivated by exposure to intense sunlight the following day. Treating on cloudy (but not rainy) days provides a similar result. Production processes that encapsulate Bt spores or toxins in a granular matrix (such as starch) or within killed cells of other bacteria also provide protection from ultraviolet radiation. Registration and sale of products containing encapsulated Bt are forthcoming.

Other bacterial insecticides:

Insecticides sold under the trade names Doom®, Japidemic®, Grub Attack®, and the generic name «milky spore disease» contain the bacteria *Bacillus popillae* and *Bacillus lentimorbus*. These bacteria cannot be grown in fermentation tanks; instead, they are "cultivated" in laboratory-reared insect larvae. Products containing *Bacillus popillae* and *Bacillus lentimorbus* can be applied to turf and "watered in" to the soil below to control the larval (grub) stage of the Japanese beetle and, less effectively, some other beetle grubs. When a susceptible grub consumes spores of these bacteria, they proliferate within it, and the grub's internal organs are liquefied and turned milky white (hence the name, milky spore disease). These symptoms develop slowly, often over a period of three to four weeks following initial infection.

Limitations

Unfortunately, *Bacillus popillae* and *Bacillus lentimorbus* offer limited usefulness because the predominant lawn grubs in this region are annual white grubs, which are larvae of beetles called chafers (genus Cyclocephala). These larvae are not susceptible (or are only slightly susceptible) to milky spore disease.

Viruses (NPV - Nuclear polyhedrosis virus)

NPV which belongs to the sub group Baculoviruses is a virus affecting insects, predominantly larval stages of moths and butterflies. NPVs are largely restricted to specific host insect.

The viruses that are nuclear polyhedrosis viruses (NPVs), in which numerous virus particles are "packaged" together in a crystalline envelope within insect cell nuclei, or granulosis viruses (GV's), in which one or two virus particles are surrounded by a granular or capsule like protein crystal found in the host cell nucleus, these groups of viruses infect caterpillars and the larval stages of sawflies. Viruses, like bacteria, must be ingested to infect insect hosts. In sawfly larvae, virus infections are limited to the gut, and disease symptoms are not as obvious as they are in caterpillars.

Caterpillars killed by viral infection appear limp and soggy. They often remain attached to foliage or twigs for several days, releasing virus particles that may be consumed by other larvae. The pathogen can be spread throughout an insect population in this way (especially when rain drops help to splash the virus particles to adjacent foliage) and by infected adult females depositing virus contaminated eggs. Dissemination of viral pathogens is deterred by exposure to direct sunlight, because direct ultraviolet radiation destroys virus particles. Although naturally occurring epidemics do control certain pests, these epidemics rarely occur before pest populations have reached out peak levels.

Several insect viruses have been developed and registered for use as insecticides. Most are specific to a single species or a small group of related forest pests, for example the gypsy moth, douglas-fir tussock moth, spruce budworn, and pine sawfly.

Limitations of viral insecticides

Each viral insecticide has a limited market. These economic factors, coupled with the fact that some virus insecticides are considerably less effective than available synthetic chemical insecticides, have limited their development. Examples of viral insecticides: Baculoviruses.

Fungi

An entomopathogenic fungus is a fungus that can act as a parasite of insects and kills seriously. The more potent fungi that can be used for insect control belongs the class imperfecti.

Fungi are an as important natural control agent that limit insect populations. Most of the species that cause insect diseases spread by means of asexual spores called conidia. Fungi do not have to be ingested to cause infections. In most instances, as fungal infections progress, infected insects are killed by fungal toxins, not by the chronic effects of parasitism. Fungi are used as insecticides in the following manner:

I. *Beauveria bassiana*: This common soil fungus has a broad host range that includes many beetles and fire ants. It infects both larvae and adults of many species.

II. *Verticillium lecanii* : This fungus (once sold under the trade name Vertelec®) has been used in greenhouses in Great Britain to control aphids and whiteflies.

III. *Nomuraea rileyi*: In soybeans (especially in the southeastern United States), naturally occurring epidemic infections of *Nomuraea rileyi* cause dramatic reductions in populations of foliage-feeding caterpillars.

IV. *Lagenidium giganteum*: This aquatic fungus is highly infectious to larvae of several mosquito genera. It cycles effectively in the aquatic environment (spores produced in infected larvae persist and infect larvae of subsequent generations), even when mosquito density is low. Its effectiveness is limited by high temperature.

V. *Hirsutella thompsonii*: Although preparations of this pathogen were once registered by the US EPA and marketed under the trade name Mycar®, it is no longer available commercially. *Hirsutella thompsonii* is a pathogen of the citrus rust mite.

Protozoa

Protozoan pathogens naturally infect a wide range of insect hosts. Although these pathogens can kill their insect hosts, many are more important for their chronic, debilitating effects. One important and common consequence of protozoan infection is a reduction in the number of offspring produced by infected insects. Although protozoan pathogens play a significant role in the natural limitation of insect populations, few appear to be suited for development as insecticides.

Species in the genera *Nosema* and *Vairimorpha* seem to offer the greatest potential for use as insecticides. Pathogens in these genera attack Lepidopteran larvae and insects in the order Orthoptera (the grasshoppers and related insects).

The one protozoan currently available in a registered insecticidal formulation is the microsporidian *Nosema locustae*, a pathogen of grasshoppers. It is sold as NOLO Bait® and Grasshopper Attack®. It is most effective when ingested by immature grasshoppers (the early nymphal stages).

Mechanism

Infections progress slowly; where the pathogen kills the grasshopper, death occurs three to six weeks after initial infection. Not all infected hoppers are killed.

Nematode

Nematodes used in insecticidal products are nearly microscopic in size. The entomogenous nematodes *Steinernema feltiae* (sometimes identified as *Neoaplectana carpocapsae*), *S. scapteriscae, S. riobravis, S. carpocapsae* and *Heterorhabditis heliothidis* are the species most commonly used in insecticidal preparations. Within each of these species, different strains exhibit differences in their abilities to infect and kill specific insects. In general, however, these nematodes infect a wide range of insects. On a worldwide basis, laboratory or field applications have been effective against over 400 pest species, including numerous beetles, fly larvae, and caterpillars

Steinernema species infect host insects by entering through body openings, the mouth, anus, and spiracles (breathing pores). *Heterorhabditis* juveniles also enter host insects through body openings, and in some instances are also able to penetrate an insect's cuticle.

Mechanism

»   If the environment is warm and moist, these nematodes complete their life cycle within the infected insect. Infective juveniles moult to form adults, and these adults produce a new generation within the same host. As the offsprings mature to the J_3 stage, they are able to leave the dead insect and seek a new host.

»   Nematodes have been used to effectively control flea larvae living in lawns and outdoor areas accessible to pets. The nematodes effectively kill flea larvae and have no toxicity to pets or humans.

»   Nematodes are not appropriate for termite control. Entomogenous nematodes are infectious to termites, but they are not likely to provide the long-term persistence needed in a termiticide.

Trade names:

Trade names include Vector®, Scanmask® and BioSafe®; some products simply use the scientific name of the nematode.

Advantages

»   The organisms used in microbial insecticides are essentially nontoxic and non-pathogenic to wildlife, humans, and other organisms not closely related to the target pest.

»   The toxic action of microbial insecticides is often specific to a single group or species of insects and this specificity means that most microbial insecticides do not directly affect beneficial insects.

»   Most microbial insecticides can be used in conjunction with synthetic chemical insecticides because in most cases the microbial product is not deactivated or damaged by residues of conventional insecticides.

»   Because their residues present no hazards to humans or other animals, microbial insecticides can be applied even when a crop is almost ready for harvest.

»   In some cases, the pathogenic microorganisms can become established in a pest population or its habitat and provide control during subsequent pest generations or seasons.

Disadvantages

»   Because a single microbial insecticide is toxic to only a specific species or group of insects, each application may control only a portion of the pests present in a field, garden, or lawn.

»   Heat, desiccation (drying out), or exposure to ultraviolet radiation reduces the effectiveness of several types of microbial insecticides.

»   Special formulation and storage procedures are necessary for some microbial pesticides. Although these procedures may complicate the production and distribution of certain products, storage requirements do not seriously limit the handling of microbial insecticides that are widely available.

»   Because several microbial insecticides are pest-specific, the potential market for these products may be limited.

Table 1. Microbial Insecticides: A summary of products and their uses.

PATHOGEN	PRODUCT NAME	HOST RANGE	USES AND COMMENTS
BACTERIA			
Bacillus thuringiensis var. *kurstaki* (*Bt*)	Bactur®, Bactospeine®, Bioworm®, Caterpillar Killer®, Dipel®, Futura®, Javelin®, SOK-Bt®, Thuricide®, Topside®, Tribactur®, Worthy Attack®	caterpillars (larvae of moths and butterflies)	Effective for foliage-feeding caterpillars (and Indian meal moth in stored grain). Deactivated rapidly in sunlight; apply in the evening or on overcast days and direct some spray to lower surfaces or leaves. Does not cycle extensively in the environment. Available as liquid concentrates, wettable powders, and ready to use dusts and granules. Active only if ingested.
Bacillus thuringiensis var. israelensis (*Bt*)	Aquabee®, Bactimos®, Gnatrol®, LarvX®, Mosquito Attack®, Skeetal®, Teknar®, Vectobac®	larvae of *Aedes* and *Psorophora* mosquitoes, black flies, and fungus gnats	Effective against larvae only. Active only if ingested. *Culex* and *Anopheles* mosquitoes are not controlled at normal application rates. Activity is reduced in highly turbid or polluted water. Does not cycle extensively in the environment. Applications generally made over wide areas by mosquito and blackfly abatement districts.
Bacillus thuringiensis var. *tenebrinos*	Foil® M-One® M-Track®, Novardo® Trident®	larvae of Colorado potato beetle, elm leaf beetle adults	Effective against Colorado potato beetle larvae and the elm leaf beetle. Like other *Bts*, it must be ingested. It is subject to breakdown in ultraviolet light and does not cycle extensively in the environment.
Bacillus thuringiensis var. *aizawai*	Certan®	wax moth caterpillars	Used only for control of was moth infestations in honeybee hives.
Bacillus popilliae and *Bacillus lentimorbus*	Doom®, Japidemic®,® Milky Spore Disease, Grub Attack®	larvae (grubs) of Japanese beetle	The main Illinois lawn grub (the annual white grub, *Cyclocephala* sp.) Is NOT susceptible to milky spore disease. The disease is very effective against Japanese beetle grubs (not a major pest in Illinois) and cycles effectively for years in the soil.
Bacillus sphaericus	Vectolex CG®, Vectolex WDG®	larvae of *Culex*, *Psorophora*, and *Culiseta* mosquitos, larvae of some *Aedes* spp.	Active only if ingested, for use against *Culex*, *Psorophora*, and *Culiseta* species; also effective against *Aedes vexans*. Remains effective in stagnant or turbid water. Commercial formulations will not cycle to infect subsequent generations.
FUNGI			
Beauveria bassiana	Botanigard®, Mycotrol®, Naturalis®	aphids, fungus gnats, mealy bugs, mites, thrips, whiteflies	Effective against several pests. High moisture requirements, lack of storage longevity, and competition with other soil microorganisms are problems that remain to be solved.
Lagenidium giganteum	Laginex®	larvae of most pest mosquito species	Effective against larvae of most pest mosquito species; remains infective in the environment through dry periods. A main drawback is its inability to survive high summertime temperatures.

Table 1. Microbial Insecticides: A summary of products and their uses.

PATHOGEN	PRODUCT NAME	HOST RANGE	USES AND COMMENTS
PROTOZOA			
Nosema locustae	NOLO Bait®, Grasshopper Attack®	European cornborer caterpillars, grasshoppers and mormon crickets	Useful for rangeland grasshopper control. Active only if ingested. Not recommended for use on a small scale, such as backyard gardens, because the disease is slow acting and grasshoppers are very mobile. Also effective against caterpillars.
VIRUSES			
Gypsy moth nuclear plyhedrosis (NPV)	Gypchek® virus	gypsy moth caterpillars	All of the viral insecticides used for control of forest pests are produced and used exclusively by the U.S. Forest Service.
Tussock moth NPV	TM Biocontrol-1®	tussock moth caterpillars	
Pine sawfly NPV	Neochek-S®	pine sawfly larvae	
Codling moth granulosis virus (GV)	(see comments)	codling moth caterpillars	Commercially produced and marketed briefly, but no longer registered or available. Future re-registration is possible. Active only if ingested. Subject to rapid breakdown in ultraviolet light.
ENTOMOGENOUS NEMATODES			
Steinernema feltiae (=*Neoaplectana carpocapsae*) *S. riobravis, S. carpocapsae* and other *Steinernema* species	Biosafe®, Ecomask®, Scanmask®, also sold generically (wholesale and retail), Vector®	larvae of a wide variety of soil-dwelling and boring insects	*Steinernema riobravis* is the main nematode species marketed retail in the U.S. Because of moisture requirements, it is effective primarily against insects in moist soils or inside plant tissues. Prolonged storage or extreme temperatures before use may kill or debilitate the nematodes. Effective in cool temperatures.
Heterorhabditis heliothidis	currently available on a wholesale basis for large scale operations	larvae of a wide variety of soil-dwelling and boring insects	Not commonly available by retail in the U.S.; this species is used more extensively in Europe. Available by wholesale or special order for research or large-scale commercial uses. Similar in use to *Steinernema* species but with some differences in host range, infectivity, and temperature requirements.
PATHOGEN			
Steinernema scapterisci	Nematac®S	late nymph and adult stages of mole crickets	*S. scapterisci* is the main nematode species marketed to target the tawny and southern mole cricket. Best applied where irrigation is available. Irrigate after application.

Insect Growth Regulators

Insect Growth Regulators (IGRs) are compounds which interfere with the growth, development and metamorphosis of insects. It basically mostly inhibits the life cycle of an insect. IGRs include synthetic analogues of insect hormones such as ecdysoids and juvenoids and non-hormonal compounds such as precocenes (Anti JH) and chitin synthesis inhibitors. IGRs are typically used as insecticides to control populations of harmful insect pests such as cockroaches and fleas.

Advantages

> » Many IGRs are lebelled reduced risk pesticides by the Environmental Protection Agency

> » Effective in minute quantities, so it is economical.

> » Target specific and so safe to natural enemies

> » Bio-degradable, non-persistent and non-polluting. Non-toxic to humans, animals and plants

> » More compatible with pest management systems that use biological controls.

> » Insects can become resistant to insecticides but they are less likely to become resistant to IGRs.

Disadvantages of IGRs

> » Kill only certain stages of pest.

> » Slow mode of action.

> » Since they are chemicals possibility of build-up of resistance.

> » Unstable in the environment.

Mechanism of Action

As an insect grows it moults, growing a new exoskeleton under its old one and then shedding the old one to allow the new one to swell to a new size and harden. IGRs work by interfering with an insect's moulting process to prevent it from reaching adulthood. They kill insects more slowly than traditional insecticides. Death typically occurs within 3 to 10 days, depending on the IGR product, the insect's life stage at the time the product is applied, and how quickly the insect develops. Some IGRs cause insects to stop feeding long before they die.

Type of IGRs

{a} Natural Hormones of Insects

I. Brain hormone: They are also called activation hormone (AH). AH is secreted by neuro secretory cells (NSC) which are neurons of central nervous system (CNS). Its role is to activate the corpora allata to produce juvenile hormone (JH).

II. Juvenile hormone: Also called neotinin. It is secreted by corpora allata which are paired glands present behind insect brain. Their role is to keep the larva in juvenile condition. JH I, JH II, JH III and JH IV have been identified in different groups of insects. The concentration of JH decreases as the larva grows and reaches pupal stage. JH I, II and IV are found in larva while JH III is found in adult insects and are important for development of ovary in adult females.

III. Ecdysone

Also called Moulting hormone (MH). Ecdysone is a steroid and is secreted by Prothoracic Glands (PTG) present near prothoracic spiracles. Moulting in insects is brought about only in the presence of ecdysone. Ecdysone level decreases and is altogether absent in adult insects.

{b} Synthetic analogues of insect hormones

I. Juvenoids (JH Mimics)

JH mimics were first identified by Williams and Slama in the year 1966. Juvenoids are synthetic analogues of Juvenile Hormone (JH). They can kill the insect in several ways:

>   by the antimetamorphic effect: forcing the larva to continue as larva.

>   by the larvicidal effect: producing dearrangements in the larval development.

>   by the ovicidal effect: rendering inviable eggs.

>   by the diapauses disrupting effect: preventing the pupa from diapauses.

Examples: Methoprene (against mosquito), Kinoprene (against whitefly and mealybug), Hydroprene (against cockroach and beetles), Fenoxycarb and Pyriproxyfen

II. Ecdysoids

These compounds are synthetic analogues of natural ecdysone. When applied in insects, kill them by formation of defective cuticle. The development processes are accelerated bypassing several normal events resulting in integument lacking scales or wax layer.

Examples : Methoxyfenozide, Tebufenazide etc.

{c} Non – hormonal compounds

I. Anti JH or Precocenes

They act by destroying corpora allata and preventing JH synthesis. When treated on immature stages of insect, they skip one or two larval instars and turn into tiny precocious adults. They can neither mate, nor oviposit and die soon.

Examples : EMD, FMev, and PB (Piperonyl Butoxide).

Chitin Synthesis Inhibitors (CSI)

Chitin synthesis inhibitors work by preventing the formation of chitin, a carbohydrate needed to form the insect's exoskeleton. With these inhibitors, an insect grows normally until it moults. The inhibitors prevent the new exoskeleton from forming properly, causing the insect to die. Death may be quick, or take up to several days depending on the insect. Chitin synthesis inhibitors can also kill eggs by disrupting normal embryonic development. These are also quicker acting but can affect predaceous insects, arthropods and even fish. Compounds include benzoyl phenyl urea pesticides.

Examples: * Diflubenzuron * Azadiracthin * Triflumuron * Hexaflumuron * Lufenuron

{d} IGRs from Neem

Leaf and seed extracts of neem which contains azadirachtin as the active ingredient, when applied topically causes growth inhibition, malformation, mortality, reduced fecundity in insects, formation of defective insects and reduction of the life span of adults, giving rise to larval-pupal, nymphal-adult, and pupal-adult intermediates.

{e} Hormone mimics from other living organisms

Ecdysoids from plants (Phytoecdysones) have been reported from plants like mulberry, ferns and conifers. Juvenoids have been reported from yeast, fungi, bacteria, protozoans, higher animals and plants

Question - Answers

SAQs

1. Who 1st coined the term microbial control and in which year?
Steinhaus in 1949

2. Which are the traits that are desirable for pathogens in microbial control?
Suitable strain, virulence, toxins, persistence.

3. Which are the non injested microbes in pathogen control?
Fungi and Nematode.

4. Bt toxin is useful against which insect orders?
Diptera, Lepidoptera, Coleoptera.

5. NPVs are effective against which insect order?
Lepidoptera (moths and butterflies).

6. Which group of fungi is effective against insects?
Deuteromycetes

7. IGRs are labelled by which international organization?
Environmental Protection Agency.

8. Which of the following is known as moulting hormone?
Ecdysone.

9. Which of the following carbohydrate compound present in insect's exoskeleton?
Chitin.

10. Which one of the following is an example of juvenoids hormone?
a. Azadiracthin, b. Fenoxycarb, c. Triflubenzuron, d. Triflumuron

Ans: c. Fenoxycarb

11. How microbial insecticides are applied?
Microbial insecticides are formulated to be applied as conventional insecticidal sprays, dusts, liquid drenches, liquid concentrates, wettable powders, or granules.

12. Which groups of viruses infect caterpillars and the larval stages of sawflies?

Nuclear Polyhedrosis Viruses (NPV's) and Granulosis Viruses (GV's).

13. How *Steinernema* species infects the host insects?

Steinernema species infect host insects by entering through body openings--the mouth, anus, and spiracles (breathing pores).

14. What are Insect Growth Regulators (IGRs)?

Insect Growth Regulators (IGRs) are compounds which interfere with the growth, development and metamorphosis of insects. IGRs include synthetic analogues of insect hormones such as ecdysoids and juvenoids and non-hormonal compounds such as precocenes (Anti JH) and chitin synthesis inhibitors.

15. What is the role of brain hormone in insects?

The role of brain hormone is to activate the corpora allata to produce juvenile hormone (JH).

16. Which group of fungi is effective against insects?

Deuteromycetes

A. Fill in the blanks

1) *Bacillus popillae* kills _____ but is not effective against the closely related annual white grubs.

Ans. Japanese beetle larvae

2) Bt toxins in a granular matrix or within killed cells of other bacteria provide protection from _____.

Ans. Ultraviolet radiation

3) Protozoan species in the genera ____offer the greatest potential for use as insecticides.

Ans. *Nosema* and *Vairimorph*

4) *Nosema locustae*, a pathogen of grasshoppers, is sold as _____.

Ans. NOLO Bait® and Grasshopper Attack®

5) Nematodes have been used to effectively control _____.

Ans. Flea larvae

6) Microbial insecticides also are known as _____.

Ans. Biological pathogens and biological control agents.

MCQs

1) Which active ingredient of neem also act as an insect growth regulator (IGR)?

(a) Azadirachtin   (b) Nimbin   (c) Salannin   (d) Gedunin

Ans. (a) Azadirachtin

2) *Lagenidium giganteum* is a

(a) Bacteria   (b) Fungi   (c) Virus   (d) Nematode

Ans. (b) Fungi

3) Which order of insect is affected by the pathogens of genera *Nosema* and *Vairimorpha*?

(a) Diptera   (b) Hymenoptera   (c) Odonata   (d) Orthoptera

Ans. (d) Orthoptera

4) Which of the following is an example of Anti JH?

(a) Methoxyfenozide   (b) Tebufenazide
(c) Piperonyl Butoxide   (d) Lufenuron

Ans. (c) Piperonyl Butoxide

True or False

1) Nematodes are appropriate for termite control.

Ans. False

2) Ecdysone are also called as Moulting hormone (MH).

Ans. True

3) IGRs are Target specific and so safe to natural enemies.

Ans. True

4) The residues of microbial insecticides are harmful to humans and other animals.

Ans. False

5) Juvenile hormone (JH) is secreted by corpora allata which are paired glands present behind insect brain.

Ans. True

Chapter - 16

Application of IPM Techniques, Integration and Case Studies

Practicable IPM Practices

1. **Cultural methods**

   a. Crop rotation (eg. Nematode and stage of insect inhabiting in soil)

   b. Trap crops (eg. Merry gold for tomato fruit fly)

   c. Summer tillage (Termite and Nematode)

   d. Altered timings (eg. Mustard aphid)

   e. Clean cultivation (eg. Brinjal fruit and shoot borer and fruit fly of cucurbits)

   f. Soil manuring and fertilization

2. **Mechanical methods**

   1. Hand picking the caterpillars

   2. Beating: Swatting housefly and mosquito

   3. Shaking the plants: Passing kerosene wetting rope across rice field to dislodge larvae of leaf folder

   4. Hooking: Iron hook is used against adult rhinoceros beetle

   5. Crushing: Bed bugs and lice

Mechanical exclusion:

Mechanical barriers prevent access of pests to hosts.

1. Wrapping the fruits: Covering with polythene bag against fruit fly.

2. Banding: Banding with grease or polythene sheets - Mango mealy bug.

3. Netting: Mosquitoes, vector control in green house.

4. Sand barrier: Protecting stored grains with a layer of sand on the top.

6. Water barrier: Ant pans for ant control.

7. Tin barrier: Coconut trees protected with tin band to prevent rat damage.

8. Electric fencing: Low voltage electric fences against rats.

Appliances in controlling the pests

1. Light traps: Most adult insects are attracted towards light in night. This principle is used to attract the insect and trapped in a mechanical device.

2. Pheromone trap: Synthetic sex pheromones are placed in traps to attract males. The rubberized septa, containing the pheromone lure are kept in traps designed especially for this purpose and used in insect monitoring / mass trapping programmes.

3. Yellow sticky trap: Whitefly, aphids, thrips prefer yellow colour. Yellow colour is painted on tin boxes and sticky material like vaseline is smeared on the surface. These insects are attracted to yellow colour and trapped on the sticky material.

4. Bait trap: Attractants placed in traps are used to attract the insect and kill them. Eg. Fruit fly.

3. Physical control

Modification of physical factors in the environment to minimize (or) prevent pest problems. Use of physical forces like temperature, moisture etc. in managing the insect pests.

A. Manipulation of temperature

1. Sun drying the seeds to kill the eggs of stored product pests.

2. Hot water treatment (50 - 55°C for 15 min) against rice white tip nematode.

3. Cold storage of fruits and vegetables to kill fruit flies (1 - 2°C for 12 - 20 days).

B. Manipulation of moisture

1. Alternate drying and wetting of rice fields against BPH.

2. Drying seeds (below 10% moisture level) affects insect development.

3. Flooding the field for the control of cutworms.

4. Biological control

The utilization of natural enemies and plant products for the regulation of pest population densities.

A. Parasitoid: A parasitoid is an insect which is parasitic on immature or adult stages of another insect. In the process of development it may slowly weaken and kill the host insect.

1. Egg parasitoid: *Trichogramma chilonis* – eggs of rice leaf folder.

2. Larval parasitoid: *Bracon hebetor* – larvae of lepidopteran pests

B. Predator: A predator is an insect which directly consumes or devours other insects as their food.

1. Lady bird beetle: feed on small soft bodied insects like aphid, mealy bug, jassid etc.

2. Lace wing bug: *Chrysoperla carnea* feeds on aphid and eggs of lepidopteran insect.

C. Microbes: Defined as control of pests by use of microorganisms like viruses, bacteria, protozoa, fungi, rickettsia and nematodes.

Viruses

Viruses coming under family *Baculoviridae* cause disease in lepidoptera larvae. eg. NPV (Nuclear Polyhedrosis Virus) e.g. HaNPV, SlNPV

Symptoms

Lepidopteran larvae become sluggish, pinkish in colour, lose appetite, body becomes fragile and rupture to release polyhedra (virus occlusion bodies). Dead larvae hang from top of plant with prolegs attached (Tree top disease)

Bacteria: They generally produce spores and also toxin (endotoxin). The endotoxin paralyses gut when ingested e.g. *Bacillus thuringiensis* is effective against lepidopteran. Commercial products - Delfin, Dipel, Thuricide

Fungi

i.   Green muscardine fungus - *Metarhizium anisopliae* attacks coconut rhinoceros beetle

ii. White muscardine fungus - *Beauveria bassiana* against lepidopteran larvae

D. Botanicals: They are usually different plant products which keep the pest infestation less from attack to target crop host either through repellency or antifeedant beahviour. eg. neem based botanicals like neem oil, neem seed kernel extract etc.

5.   Chemical Control: Management of insect pests using chemical pesticides is termed as chemical control.

Practical exercise on IPM of transplanted Rice

a. Treatment = 7 with replication 3

1.   Cultural control (Group – 1)

i.   Manual weeding for one time at 30-35 DAT

ii. Soil manuring and fertilization (NPK at recommended dose)

iii. No pesticide application

2.   Mechanical control (Group – 2)

i.   Hand picking the caterpillars

ii. Passing kerosene wetting rope across rice field to dislodge larvae of leaf folder

iii. No weeding

iv. No manuring and fertilization

v.   No pesticide

3. Physical control (Group – 3)

i.   Alternate drying and wetting of rice fields against BPH.

ii. No other treatments taken for cultural and mechanical control.

4. Biological control (Group – 4)

i.   Need based application of neem oil

ii.   No other treatments taken for cultural, mechanical and physical control.

5. Chemical Control (Group – 5)

i.   Need based application of chemical insecticide

ii.   No other treatments taken for cultural, mechanical, physical and biological control

6. IPM without chemical control (Group – 6)

i.   Soil manuring and fertilization (NPK at recommended dose).

ii.   Manual weeding for one time at 30-35 DAT.

iii.   Hand picking the caterpillars

iv.   Passing kerosene wetting rope across rice field to dislodge larvae of leaf folder

v.   Alternate drying and wetting of rice fields against BPH.

vi.   Need based application of neem oil.

7. IPM (group – 7)

i.   Soil manuring and fertilization (NPK at recommended dose).

ii.   Manual weeding for one time at 30-35 DAT.

iii.   Hand picking the caterpillars

iv.   Passing kerosene wetting rope across rice field to dislodge larvae of leaf folder

v.   Alternate drying and wetting of rice fields against BPH.

vi.   Need based application of neem oil and chemical pesticide

Observing parameters : Pest infestation data, yield and economics.

IPM Case Studies

How to present a case study

- » Background
- » Intervention rationale
- » Objectives
- » Methodologies used
- » Result
- » Lessons learned

Background

- » Information on the crop
- » Information on the pest
- » Management scenario

Intervention Rationale

Highlighting the proper justification for the taken study

Objective

- » To compare IPM technology of paddy (against stem borer) with other components of IPM with respect to pest damage and economic return

Methodologies used

- » Mention the process of implementation highlighting resources and materials have been utilized
- » Selecting parameters for conducting the study

Result

- » Details presentation of data in tabular or graphical form.
- » Descriptive write up of the tabulated or graphical data for clear understanding of the study

Lessons learned:

> » Drawing conclusion having a concrete recommendation for further study or real application

Title of Case Study: Case study of IPM and its individual components against stem borer (*Scirpophaga incertulas*) in boro paddy.

a. **Background:** Rice is the principle crop in Burdwan district of West Bengal. It is cultivated in large areas of the district mainly as kharif and boro paddy. In boro season, the crop is attacked by yellow stem borer, *Scirpophaga incertulas* which is recognized as a major insect. Farmers rely on the frequent use of conventional insecticides for its management without considering threshold damage.

b. **Intervention rationale:** Dependence on injudicious sole application of conventional chemical insecticides for management of paddy stem borer has produced negative impact on natural enemies, human health, soil and environment. There is also a chance of insecticide resistance development for the same. Emphasis is to be given to replace hazardous conventional insecticides by eco-friendly novel insecticides and bio-pesticides. Knowledge on damage threshold (ETL) of paddy stem borer is also important to reduce the cost of cultivation. To deal with any insect problem, IPM could be treated as a best alternative way where its different components are intelligently integrated in an economical way. It is imperative to develop environment friendly, economically sustain and area specific applicable plant protection schedules.

c. **Objectives:** The said case study has been conducted by the students in groups to gain practical learning on IPM of boro paddy. The main objectives of this study:

1) To compare IPM technology of paddy (against stem borer) with other components of IPM with respect to pest damage and economic return

d. **Methodologies used**

1. Students (32) were divided into eight (8) different groups.

2. Each group was given with responsibility to execute a trial treatment against paddy stem borer (Gr.1: Cultural..........) with three replications for each.

3. Recording data on stem borer damage in paddy (% number white ear head), marketable paddy yield and economic benefit at harvest.

e: Result

Table 1: Effect of treatment schedules on damage and economics of paddy by stem borer

Treatment	Percent (%) number of white ear head	Damage grain yield in kg/ha	Marketable yield in kg/ha	Extra yield in kg/ha w.r.t un-untreated control	Extra gross return in Rs./ha	Cost for treatment in Rs./ha	Net return in Rs./ha	B:C ratio
Cultural								
....								
....								

f. Lessons learned:

Sample of IPM case study: Spillover effect of IPM technology among fellow farmers

Mrs. Radhika Devi, 35 is a female non participant IPM farmer from village Kultali. She is married having 2 children. She is literate with primary education. Her husband works in farm and occasionally goes for wage labour in nearby villages. Her family is the commercial vegetable grower. They are small farmers with only 1.5 bighs of land. They grow vegetables crop all the year round. All the lands are cultivable and irrigated. She earned a net profit of Rs. 1,50,000/year. But, this was not enough to maintain her family. She was not having any knowledge on IPM of vegetable. She managed her vegetables crops from pest attack as per suggestion from local dealer only. So cost for plant protection was very high with less profit.

Objective: To know the spillover effects of IPM technology in project sites

Rationale: As IPM has been considered to be sustainable approach to healthy crop production and pest management that builds social capital of farmer groups intentionally or not, non participant of IPM FFS can be affected by programs, and the spillover effects should be taken into account, the survey of spillover effects of National IPM Programme is necessary.

Procedure: Site observation, interview technique

She had immense interest to be a part of IPM groups, both husband and wife were absent during the farmer selection process of FFS. She was always in the field when her farm neighbours (who were IPM FFS participants) carry out any activities. She asked about some of the IPM technologies from her neighbours. With the success of IPM farmers in increasing profit, her family started to believe in IPM techniques. Her family started adoption of the most of IPM technologies to grow vegetables.

Result: She is now able to identify quality seed, variety and manage need based application dose of chemicals. She can easily identify harmful and beneficial insects present in crop. She praises herself that she is able to practice the IPM technologies in spite of her absence in IPM FFS conducted in her village. She encourages her husband to use botanical pesticides.

She shared that she uses pesticides to protect crops from losses. However, she argued that they never got premium price for using IPM technologies. Now her net profit has been increased by Rs. 50,000 by adoption of IPM practice.

Lessons learned

The case study revealed that non participants of IPM FFS can also gain the knowledge over time when they perceive the technology is important and profitable to them in short or long run. Cluster approach of site selection and year long FFS is a major factor for making behavioural and cultural change in crop production and pest management.

Question - Answer

Fill in the blanks

a) Butterflies are picked out by using_____
Ans: Hand nets

b) Crop rotation is a _____ method of IPM.
Ans: Cultural

c) Refrigeration at _____ of edibles including dry fruits will kill the insect
Ans: 5-10°c

d) Early sowing of _____ Jowar to escape from attack of Jowar shoot fly.
Ans: Kharif

e) IPM Farmer Field Schools were started in _____in Indonesia to reduce farmer reliance on pesticides in _____.
Ans: 1989, rice

SAQs

5. What are ecosystem services and their significance?

Rivers, streams and wetlands provide people with a wide range of benefits often referred to as "Ecosystem Services". Ecosystem services are classified into four broad categories:

i.   Provisioning services - the goods or products obtained from ecosystems.

ii. Regulating services - the benefits obtained from an ecosystem's control of natural processes.

iii. Cultural services - the non material benefits people obtain from ecosystem services.

iv. Supporting services - the underlying processes that are necessary for the production of all other ecosystem services. Ecosystem Services approach provides a frame work by which ecosystem services are integrated into public and private decision making. Its implementation typically incorporates a variety of methods includes

  1. Ecosystem service dependency and impact assessment

  2. Valuation

  3. Scenarios

Policies and other interventions targeted at sustaining ecosystem services:

(1) Transformations of natural assets into products valued economically.

(2) Transformations of the by-products - ecosystem services back into natural assets.

(3) Internal transformations among natural assets to maintain those assets.

5.   What is pest surveillance? What are its objectives?

Pest Surveillance is the constant watch on population dynamics of pest, its incidence and damage on each crop at fixed intervals, to fore-warn farmers to take-up timely crop protection measures. It can be done using the light traps, pheromone traps, food traps, attractants, pitfall traps (for soil insects), field scouting etc.

Objectives of pest surveillance:

- To know existing and new pest species.

- To assess pest population and damage at different growth stage of crop.

- To study the influence of weather parameters on pest.

- To study changing pest status (minor to major)

- To assess natural enemies and their influence on pests.

- To assess the effect of new cropping pattern and varieties on pest.

6.   What are the types of pests on the basis of occurrence? Give examples.

On the basis of occurrence, pests are categorised as:

1.   Regular pest: Frequently occurs on crop, close association e.g. Rice stem borer, Brinjal fruit borer etc.

2.   Occasional pest: Infrequently occurs, no close association e.g. Caseworm on rice, Mango stem borer etc.

3.   Seasonal pest: Occurs during a particular season every year e.g. Red hairy caterpillar on groundnut, Mango hoppers etc.

4.   Persistent pests: Occurs on the crop throughout the year and is difficult to control e.g. Chilli thrips, mealy bug on guava etc.

5.   Exotic pest: Non-native or non-indegenous pests which are known to occur in the state or country. e.g. Cotton mealy bug

7. Write the importance of IPM in brief.

1. To reduce pest status below economic injury level. Complete elimination of pest is not the objective. 2. To manage insects by not only killing them but by preventing feeding, multiplication and dispersal. 3. To use eco-friendly methods, which will maintain quality of environment (air, water, wild life and plant life). 4. To make maximum use of natural mortality factors, apply control measures only when needed. 5. To use component in sustainable crop production

8. What is AESA? What are the components of AESA?

Agro Ecosystem Analysis (AESA) is an approach which can be gainfully employed by extension functionaries and farmers to analyse field situations with regard to pests, defenders, soil conditions, plant health, the influence of climatic factors and their inter relationship for growing healthy crop. Basic components of AESA are: i. Plants health at different stages and monitor symptoms of diseases and nematodes. ii. Built-in-compensation abilities of the plants. iii. Pest and defender population dynamics. iv. Soil conditions. v. Climatic factors. vi. Farmers past experience.

9. Define IPM

It is the pest management system in context of associated environment and population dynamics of pest species utilize all the suitable techniques and methods in as compatible manner as possible and maintains pest populations at level below those causing economic injury.

10. In which year IPM Farmer Field Schools were started in Indonesia?

In Indonesia IPM Farmer Field Schools were started in 1989.

11. Give two examples of Cultural method of pest management.

Tillage operation and crop rotation are two cultural methods.

12. Define trap crop?

The trap crops are those crops which are used to control the pest of main crop. Trap crop plants are harvested early or used as a fodder.

13. Give one demerit of physical method of pest management?

These methods are time consuming & costly.

14. What is electric trap?

Live metal screen on which birds & insects are electrocuted.

15. What is a predator?

Predators are free living organisms that feed on living insects & consume more than one individual during their lifespan. They attack on prey at larval & adult stage. Eg. Lady Bird beetle

State true or false

1. Okra followed by cotton decreases the pest attack

Ans: False.

2. Physical methods of pest management is time consuming

Ans: False.

3. Cultural method increase the production cost.

Ans: False.

4. Cotton and oilseeds are a commercial crops

Ans: True

References

For the edition of this book the author has used the following sources

1	Fundamentals of Agricultural Entomology	P.K. Sehgal
2	Principles Of Applied Entomology	K.N. Ragumoorthi, V. Balasubramani, N. Natarajan, M.R. Srinivasan
3	Insect Pests of Horticultural Crops And Its Management	K. Senguttuvan
4	Insect Pests of Field Crops and their Management	M.R. Srinivasan and T. M Kishan
5	A sampling fork for estimating populations of small arthropods.	C. F. Henderson
6	Fundamentals of Entomology	L. C. Patel
7	A textbook of Applied Entomology Concepts in Pest Management (Vol-I)	K.P. Srivastava, G.S Dhaliwal and G. Singh
8	Principles of Applied Entomology	K.N. Ragumoorthi, M.R. Srinivasan, V. Balasubramani and N. Natarajan
9	Textbook of Introductory Plant Nematology	R.K. Walia and H.K. Bajaj
10	Internet (Various websites and Wikipedia)	www.google.com, www.farmer.gov.in, www.researchgate.net, www.wikipedia.org, www.slideshare.net, www.krishikanti.net, www.ppqs.gov.in, https://agritech.tnau.ac.in